Environmental Interactions of Clays

Springer-Verlag Berlin Heidelberg GmbH

A. Parker · J.E. Rae (Eds.)

Environmental Interactions of Clays

Clays and the Environment

With 74 Figures and 11 Tables

Springer

Dr. Andrew Parker
Dr. Joy E. Rae
The University of Reading
Postgraduate Research Institute for Sedimentology
PO Box 227
Whiteknights
Reading RG6 6AB
United Kingdom

ISBN 978-3-642-08208-5

Library of Congress Cataloging-in-Publication Data

Environmental interactions of clays / A. Parker, J.E. Rae (eds.).
p. cm. – (Clays and the environment)
Includes bibliographical references and index.
ISBN 978-3-642-08208-5 ISBN 978-3-662-03651-8 (eBook)
DOI 10.1007/978-3-662-03651-8

1. Clay minerals. 2. Clay. 3. Clay minerals – Environmental aspects. 4. Clay – Environmental apsects. I. Parker, A. (Andrew), 1941– . II. Rae, J.E. (Joy E.) III. Series.
QE389.625.E34 1998
552'.5–dc21 97–46101

Originally published by Springer-Verlag Berlin Heidelberg New York in 1998
Softcover reprint of the hardcover 1st edition 1998

Coverdesign: Erich Kirchner, Heidelberg
Typesetting: Fotosatz-Service Köhler OHG, Würzburg

SPIN: 10467937 32/3020-543210 – Printed on acid-free paper

Preface

There is a growing awareness of the key role played by clays in environmental studies. Natural assemblages of clays frequently control the behaviour of pollutants in the environment and in some instances they may even be the pollutants themselves. Clays may also be added to the environment for remediation purposes to act as adsorbents and/or catalysts. The importance of clays in environmental research is reflected in the increasing number of meetings which address this topic. Indeed, the 10th International Clay Conference in Australia in 1993 adopted the title 'Clays Controlling the Environment'.

This volume is the second in the Clays and the Environment series. The first volume deals with the occurrence of clays in nature and lays the foundation for a fuller understanding of the environmental interactions of clays which are considered here. The idea for the text arose from discussions with B. Velde and there was a meeting at Reading in 1996 of the same name, in which the chapter authors were all keynote speakers. The volume editors would like to thank B. Velde for his inspiration and encouragement and their friends and fellow scientists for their contributions.

A. Parker and J. Rae

Contents

6 Clay Barriers in Landfills
J. Arch

Contributors

Dr. Jonathan Arch
Aspinwall & Company Ltd., Walford Manor, Baschurch,
Shrewsbury SY4 2HH, UK, E-mail: jonathan.arch@aspinwall.co.uk

Dr. J.B. Dixon
Department of Soil and Crop Sciences, Texas A & M University,
College Station, Texas 77843-2474, USA, E-mail: j-dixon@tamu.edu

A.R. Gibbs
Consultant Histopathologist, Llandough Hospital, Llandough, Penarth,
South Glam CF64 2XX, UK

Dr. William A. House
Institute of Freshwater Ecology, River Laboratory, East Stoke, Wareham,
Dorset BH20 6BB, UK, E-mail: WAH@wpo.nerc.ac.uk

Dr. T.A. Jackson
Aquatic Ecosystem Restoration Branch, National Water Research Institute,
PO Box 5050, Burlington, Ontario L7R 4A6, Canada,
E-mail: t.a.jackson@cciw.ca

K. McConnochie
Tyle House, Llanmaes, Llantwit Major, Cardiff CF61 2XZ, UK

Dr. Andrew Parker
The University of Reading, Postgraduate Research Institute
for Sedimentology, PO Box 227, Whiteknights, Reading RG6 6AB, UK,
E-mail: A.Parker@reading.ac.uk

Prof. F.D. Pooley
University of Wales, School of Engineering,
Division of Materials & Minerals, PO Box 685, Newport Road,
Cardiff CF2 3TH, UK, E-mail: ball@cardiff.ac.uk

Dr. R. Pusch
Clay Technology AB, Ideon Research Center, S-22370 Lund, Sweden,
E-mail: clay@clay-tech.se

Dr. Joy E. Rae
The University of Reading, Postgraduate Research Institute
for Sedimentology, PO Box 227, Whiteknights, Reading RG6 6AB, UK,
E-mail: J.E.Rae@reading.ac.uk

J.C. Wagner
'Foxstones', 59 Coombe Valley Road, Preston, Weymouth,
Dorset DT3 6NL, UK

Chapter 1

Environmental Interactions of Clays

J. Rae and A. Parker

1.1 Introduction: Clays in Environmental Studies

Think of a list of earth-surface environmental problems which do not involve clays and it probably will not be very long. Clays play a significant role in a diverse range of environmental problems and an awareness of this role is growing.

This volume is the second in the Clays and the Environment series. The first volume, *Origin and Mineralogy of Clays* (Velde 1995) is an introduction for environmentalists to clay minerals and clay mineralogy which lays the foundation for what follows here. This volume considers some specific examples of the role of clays in environmental studies. Each chapter that follows deals with a specific medium or a particular problem (transport of radionuclides, soils, micro-organic pollutants, trace elements, landfills and mineral dusts).

It is clearly impossible to cover the full spectrum of environmental interactions of clays in one volume: our aim therefore is to illustrate the importance of clays in a wide range of environmental studies through the use of some specific examples.

1.1.1 Terminology

The most common mineralogical tool in the nineteenth century was the petrographic microscope and the smallest particle which could be distinguished was 2 µm in diameter. Thus clay minerals were originally defined as the submicroscopic but crystallized particles with diameters less than 2 µm. It is well known now, of course, that particles in this size range can include quartz, carbonates, metal oxides and other minerals in addition to clay minerals. The term clay mineral is now normally reserved for specific fine-grained layer-lattice aluminosilicates, or groups of related minerals of this type, such as kaolinite. The authors in this volume have interpreted *Clays* in a variety of ways: <2 µm particles and clay minerals (Ch. 1); smectite (Ch. 2); <2 µm particles (Ch. 3); clay minerals (Ch. 4); colloidal minerals (Ch. 5); lithologic unit (Ch. 6); and clay minerals (Ch. 7).

1.1.2 The Environmentally Significant Properties of Clays

The characteristics of clays which are used to classify them are swelling properties and crystallographic repeat units of the layer structure (see Velde 1992, 1995). The role of these two key properties in relation to classification is shown in Table 1.1 and forms the starting point for the discussion which follows.

1.1.2.1 *Particle Size and Surface Area*

Clays are by definition of fine grain size; this results in clays remaining in suspension in aqueous environments for relatively long periods of time. The significance of this suspension time relates to the enhanced potential for significant transport of pollutants associated with clay particles. A

Table 1.1. Outline of classification of clay minerals in relation to the layer distance method and swelling properties. Mixed-layer minerals and sepiolite-palygorskite minerals are also shown

Non-swelling	7 Å minerals, e.g. *kaolinite* 14 Å minerals, e.g. *chlorite* 10 Å minerals, (neutral), e.g. *talc* 10 Å minerals, (high charge), e.g. *illite*	No water sorption
	Sepiolite-palygorskite minerals, e.g. *palygorskite* (not sheet silicates in the strict sense)	Water sorption
	NB Contain chemically-absorbed water in the structure, similar to smectites	
Swelling	Mixed-layer minerals, e.g. corrensite	No water sorption
	10 Å minerals (low charge) smectites, e.g. *montmorillonite*	Water sorption

further and most important consequence of small particle size is the resulting high ratio of surface area to volume. It is well known that increasing the surface area of particles will increase the potential for adsorption of pollutants (see Sect. 1.1.2.4).

1.1.2.2
Swelling Properties

Low charge on a 10 Å structure allows hydrated/polar ions to be absorbed (i. e. inserted between the layers). This process changes the average interlayer distance, the mineral then swells and the interlayer distance increases to 12.5 Å (one water layer) or 15.2 Å (two water layers). This is important in environmental problems in which an effective 'seal' is required. Obvious applications are in the disposal of nuclear waste (Chap. 2) and in landfilling (Chap. 6).

1.1.2.3
Mineral Shape

Clay minerals are characteristically flake, lath or needle-like in shape (Fig. 1.1). The importance of shape is three-fold. Firstly the needle-like shape is important in relation to lung retention (Chap. 7). Secondly the 'flat' shape of many clay particles is a contributory factor in allowing clays to remain in suspension in natural waters for relatively long periods. Thirdly, surface area is related to shape such that sheet > lath > needle.

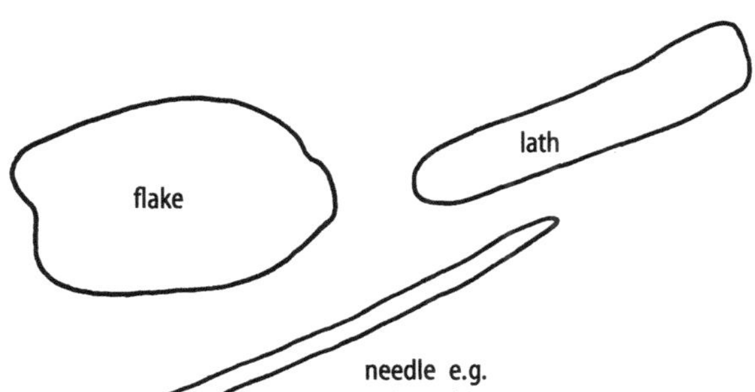

Fig. 1.1. Typical clay particle shapes. (After Velde 1995)

1.1.2.4
Surface Charge

A particularly important property of clays is the chemical activity of their surfaces which arises due to surface charge. This charge is caused by a combination of broken bonds, surface-growth defects and cation substitutions in the lattice. The net effect is to produce a negatively charged surface. This can have two important consequences for environmental studies. Firstly, when riverine clays meet saline solutions, as for example in estuarine systems, charge reversal occurs and particles form aggregates (a process referred to as flocculation). When the aggregates become large (and dense) enough to settle out of suspension, large-scale deposition of clays can occur. Siltation of inland waterways can result, in which case the clays themselves can be said to be the pollutants. Further, any anthropogenic pollutants associated with the clays will accumulate in the area of flocculation.

Another very important effect of surface charge relates to the adsorption and absorption of pollutants. Absorption is the process of accumulation of chemical species at an interface: in our case the clay surface. Adsorption is the incorporation of pollutants within the (clay) mineral (Fig. 1.2). Although both adsorption and absorption are complex processes (Parks 1975), a simplified diagrammatic representation is possible. Figure 1.3 illustrates a negatively charged clay surface in which the charge is neutralised by positive ions from the aqueous solution in contact. The charged layer on the surface of the clay is referred to as the *fixed layer*. The ions in the solution with opposite charges to the clay surface and their co-ions (with the same charge as the clay surface) are called the *diffuse or Gouy layer*. The ions in the Gouy layer are adsorbed, but can be released (exchanged for other ions), if the composition of the bulk solution changes. If the adsorbed ions are strongly associated with the clay surface (called *specific adsorption*), the fixed layer incorporates these ions. The Gouy layer will now have an excess of negative ions.

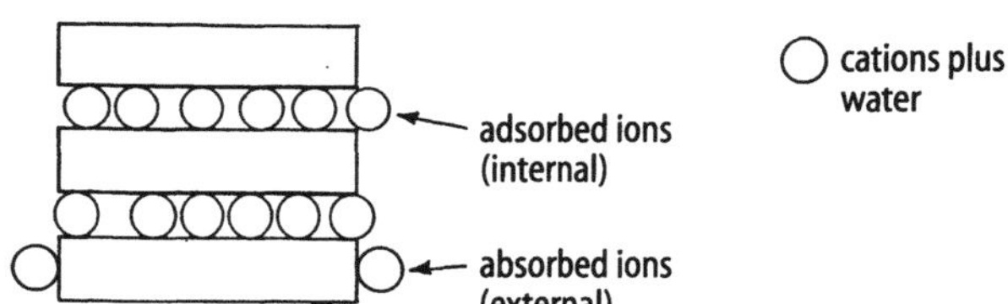

Fig. 1.2. Absorbed and adsorbed ions on clay. Adsorbed ions are found within the structure (between the layers of the mineral structure) whereas the absorbed ions are found on the surface of the particle. (After Velde 1995)

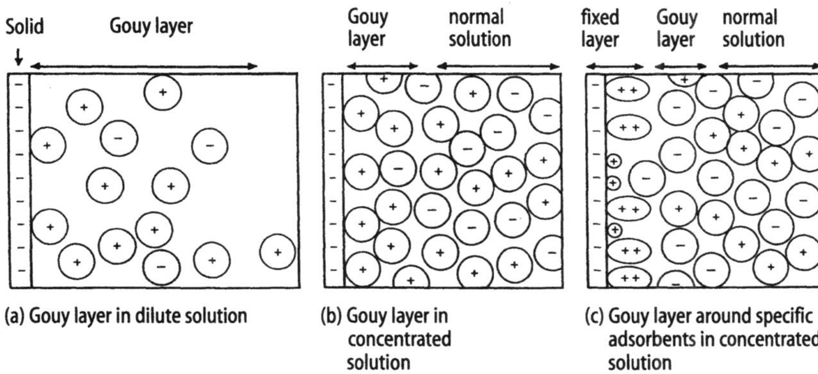

Fig. 1.3. Schematic illustration of the adsorption process. (After Fairchild et al. 1988)

The significance of the adsorption and absorption of pollutants in environmental studies cannot be overemphasised. Examples of this significance are given in Chapters 3, 4, 5 and 6. The property of adsorbing and absorbing ionic species in solution is called the cation exchange capacity (CEC). The latter is measured in terms of the total number of charged ions which can be fixed onto the clay surface, i.e. the number of moles of ionic charge fixed into 100 g of dry clay. Values are expressed in milli-equivalents of charge (moles)/100 g.

1.2 The Scope of Environmental Studies Involving Clays

As we have discussed already, there are few environmental studies in which clay particles have no role to play. Some of these roles are summarised in Fig. 1.4 with examples drawn from this book. Other important problems in which clays feature include contaminated land studies, the transport and reactivity of nutrients and the behaviour of ^{137}Cs in upland soils following the Chernobyl nuclear disaster of 1986. Some new problems are only just beginning to emerge, for example the role of clays in transporting viruses in coastal waters and in protecting absorbed viruses from degradation by sunlight. There will no doubt be an endless plethora of new and in some cases currently unimagined environmental problems in years to come. One thing is certain: the role of clays will continue to be of utmost importance.

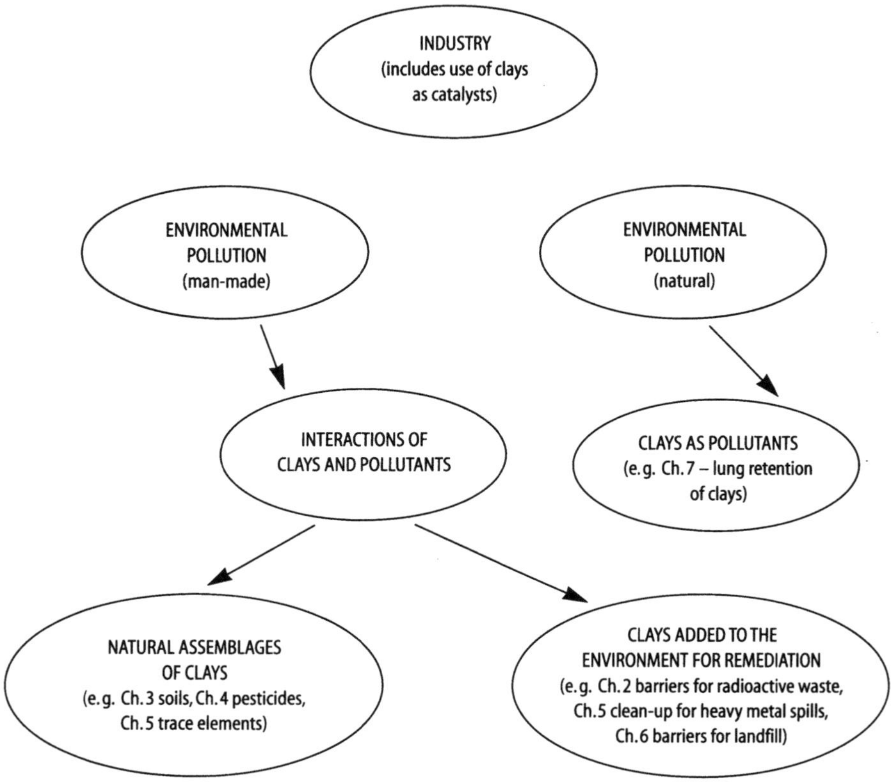

Fig. 1.4. The role of clays in environmental studies, with examples drawn from this volume

References

Fairchild I, Henry G, Quest M, Tucker M (1988) Chemical analysis of sedimentary rocks, Chapt. 9. In: Tucker M (ed) Techniques in sedimentology. Blackwell, Oxford, pp 274–354

Parks GA (1975) Adsorption in the marine environment. In: Reby JP, Skirrow G, Chemical oceanography, I. Academic Press, London, pp 241–308

Velde B (1992) Introduction to clay minerals. Chapman and Hall, London

Velde B (1995) Origin and mineralogy of clays. Springer, Berlin Heidelberg New York

Chapter 2

Transport of Radionuclides in Smectite Clay

R. Pusch

2.1 Introduction

Smectite clay is used for sealing purposes in repositories for radioactive waste because of its excellent isolating properties. Typical examples are grouts injected into rock fractures for sealing the host rock, and blocks of highly compacted clay powder for plugging shafts and tunnels and for embedding metal containers with waste. The most important clay materials are bentonites, which usually have the smectite species montmorillonite as the major clay mineral. Such clays have a low hydraulic conductivity and a swelling potential that makes them fill up and seal off the space in which they are placed. They also have a significant cation exchange capacity and, at least for some ion species, a low diffusion transport capacity. The migration of radionuclides is of course of great importance for the performance of repositories for high-level radioactive waste, a matter that will be dealt with here. This chapter starts by describing the nature of smectitic clays and by giving examples of bentonite seals, and proceeds with explaining microstructural processes that control physical properties like hydraulic and gas conductivities and swelling ability. The chapter ends by outlining the basics of the longevity of smectite clay materials and by describing the performance of smectitic seals under repository conditions.

2.2 Nature of Smectitic Clay

2.2.1 Internal and External Voids

The lattice constitution of smectite crystals is similar to that of the micas but without potassium ions as ligands, and the individual lamellae are

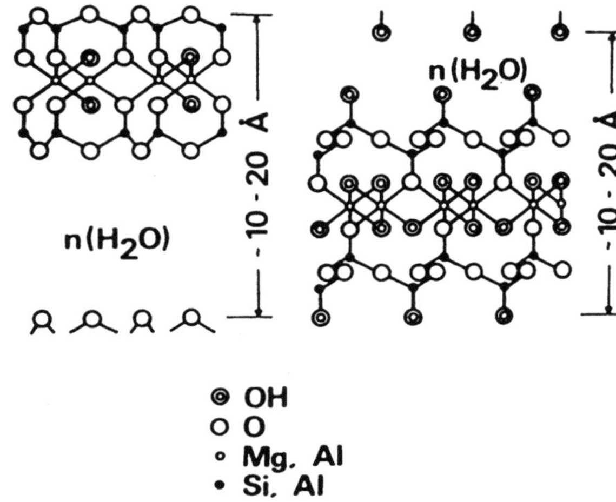

Fig. 2.1. Two possible forms of the montmorillonite lattice. *Left* Hofmann/Endell/Wilms' model (HEW) with oxygens exposed in the interlamellar space. *Right* Edelman/Favejee's model (EF) with hydroxyls protruding into the interlamellar space

therefore easily separated in conjunction with uptake of water molecules in interlamellar positions (Fig. 2.1).

The HEW version is assumed to be valid for significantly higher temperatures than room temperature and when bi- or polyvalent ions occupy the interlamellar space for compensating the negative net lattice charge (Pusch 1994a). The EF condition may prevail at low temperatures and when Li and Na are in these positions. The interlamellar cations, which are exchangeable to an extent that is manifested by the high cation exchange capacity (60–100 mEq/100 g), hold the lamellae together.

2.2.2 Interlamellar (Internal) and External Water

The interlamellar space has a strong potential for water uptake because of the hydration power of the cations or possibly, for Li and Na in interlamellar positions, because of establishment of a water lattice directly on the exposed hydroxyls (Pusch 1994a). If a stack of montmorillonite lamellae, commonly termed flakes, is free to absorb water, it is concluded to have three interlamellar hydrate layers established when Li or Na are adsorbed cations, while only two are formed if Ca and most radionuclides are adsorbed.

The interlamellar or "internal" water is largely immobile under ordinary hydraulic gradients and since it forms a larger part of the total water content in Li and Na montmorillonite than when Ca or most positively charged radionuclides are adsorbed, the swelling potential and ability to self-seal is higher in the former case (Fig. 2.2).

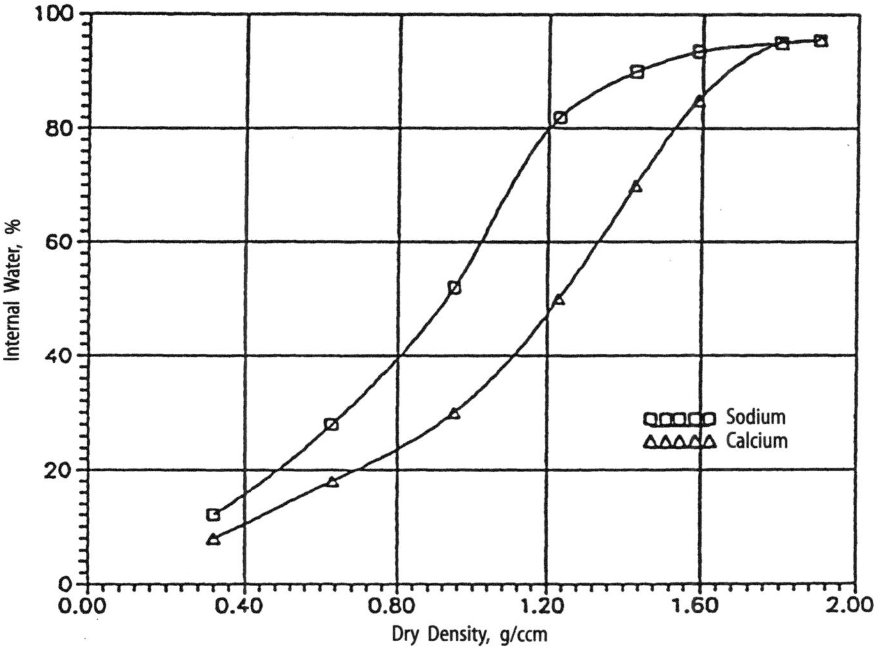

Fig. 2.2. Internal water in percent of total water content for Na and Ca montmorillonite. (Pusch 1994a)

As indicated by the diagram in Fig. 2.2, the internal, i.e. interlamellar space, represents only part of the total void system. The rest is made up of differently sized openings, ranging from the narrow space between adjacent stacks of flakes where electrical double-layers control the interparticle force field, to larger permeable voids. The water in these openings is in its normal physical condition and flows when a hydraulic gradient is applied. Figure 2.3 shows a micrograph of a system of integrated, fully expanded stacks of flakes and Fig. 2.4 a generalized sketch of the microstructure of natural smectite-rich clay.

2.3 Seals of Bentonite

2.3.1 Preparation

Figure 2.5 gives examples of how precompacted blocks of bentonite can be used to embed and isolate radioactive waste containers. The blocks are prepared by compacting commercial bentonite powder “granules” with

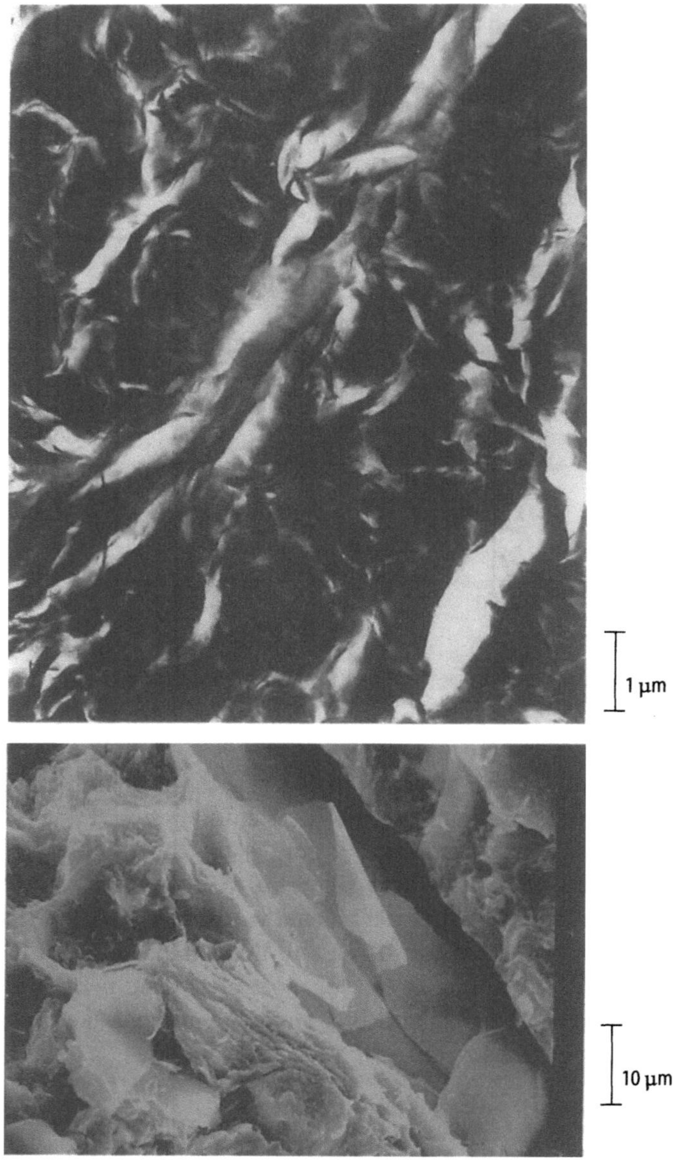

Fig. 2.3. Electron images of montmorillonite clay. *Above* Transmission micrograph of a few hundred Å thick section of polymer-embedded Triassic clay rich in montmorillonite. *Below* Scanning micrograph of freeze-dried, soft montmorillonite clay with cardhouse structure

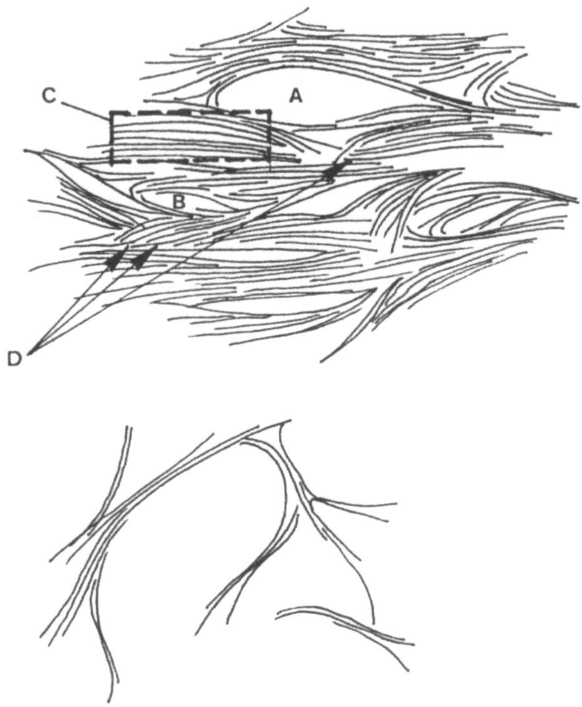

Fig. 2.4. Schematic picture of the microstructure of smectite clay. *Above* Dense clay. *A* Large, more or less continuous void. *B* Small, isolated void. *C* Interlamellar spaces. *D* Contact region of interacting stacks with electrical double-layers. *Below* Soft clay with open network of thin stacks of flakes. (Pusch 1994a)

Na or Ca as the major adsorbed cation (Pusch 1994a). With the common water content of 10–15% and with a compaction pressure of 100 MPa, the dry density (ratio of the weight of solid substance and volume including voids) is about 1.7 g/cm^3 and the degree of water saturation about 50%. Compaction of wetted powder can give blocks with up to 90% saturation. This raises the thermal conductivity, which is valuable for transferring heat that emanates from the radioactive decay to the rock, so that the temperature in the clay will not be critically high.

2.3.2 Multiple Performance

Once the blocks have absorbed water from the rock to yield complete saturation, the joints between them disappear and the blocks homogenize and come into tight contact with both the canisters and the surrounding rock. The artificially applied clay thereby turns into a dense, almost impermeable mass resembling a homogeneous bentonite bed in nature with the following properties (Pusch 1994a):

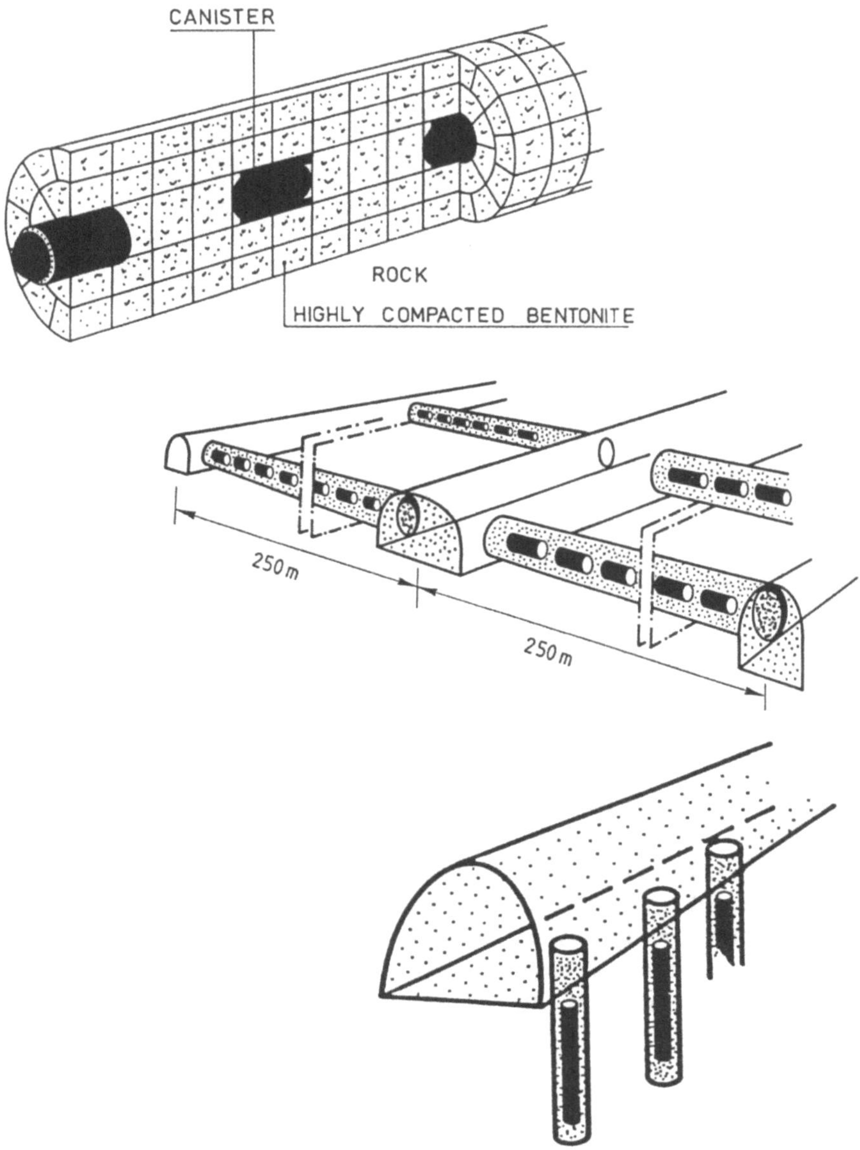

Fig. 2.5. Examples of the use of bentonite blocks in proposed repository concepts. *Upper left* Block-embedded canisters in long tunnels. *Lower right* Concept of the Swedish Nuclear Fuel and Waste Company (KBS3) with holes drilled from the floor of tunnels at 500 m depth. (Pusch 1994 a)

1. Sufficient bearing capacity to carry the heavy canisters without significant movement.
2. Sufficient ductility to prevent rock movements damaging the canisters.
3. Sufficiently low hydraulic conductivity to prevent groundwater flow near the canisters.
4. Sufficiently narrow pore system to filter radionuclide-bearing colloids.
5. Sufficient gas conductivity to release gas produced by canister corrosion.
6. Sufficient heat conductivity to transfer heat from the waste to the rock.
7. Sufficient longevity to perform acceptably for hundreds of thousands of years.

Most of these abilities are due to the crystal lattice constitution, hydration power and microstructure of smectitic clays, which will all be discussed here.

2.3.3 Clay Microstructure

The rather homogeneous microstructural pattern shown in Fig. 2.4 is not representative of compacted bentonite blocks. Instead it has a granular character inherited from the compacted powder mass as shown in Fig. 2.6. The picture represents the stage when the voids between the granules, which consist of large numbers of more or less aligned stacks of flakes, have been filled with water that is sucked up by capillary and osmotic tension forces from the surroundings. The saturation, which implies dissolution of enclosed air, is accompanied by expansion of the granules and spontaneous exfoliation of stacks of flakes from them. These stacks rearrange and coagulate to form clay gels with a density that is a function of the bulk density and of the ionic strength of the porewater as indicated in Fig. 2.6.

The bulk properties, like the gas and hydraulic conductivities and the ion diffusivity, are determined by the pore size and mechanical strength of the gel fillings. Modelling has been made of their geometrical features, assuming them to consist of regular orthogonal cardhouse stack arrangements, and it has led to microstructural parameter values of the type shown in Tables 2.1 and 2.2 (Pusch et al. 1990). Using such data and generalising the flow passages to form three-dimensional channels with regularly varying cross section, the calculated bulk hydraulic conductivity of montmorillonite clay saturated with Na and Ca has been found to agree very well with what is actually recorded.

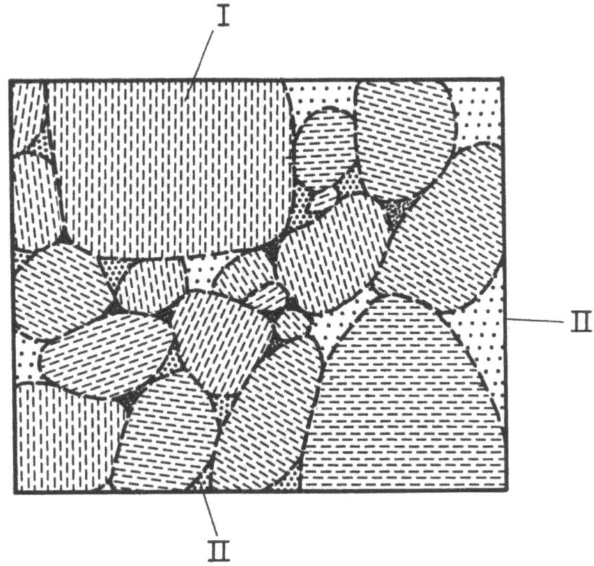

Fig. 2.6. Schematic picture of matured microstructure of compacted bentonite powder. I Dense, expanded granule. II Clay gels formed in the voids between the granules

Table 2.1. Microstructural data of compacted bentonite (D_e is the diameter of the external voids)

Bulk density (g/cm^3)	Gel density, g/cm^3		
	D_e = 1–5 mm	D_e = 5–20 mm	D_e = 20–50 mm
2.13	1.3	–	–
1.85	1.3	1.2	–
1.57	1.4	1.2	1.1

Table 2.2. Void space in gel fillings

Gel density (g/cm^3)	Free space, Å	
	Na state	Ca state
1.3	340	700
1.2	570	1100
1.1	1230	10000

2.3.4
Numerical Modelling of the "Maturation" Process

Naturally, the maturation of bentonite seals by hydration and expansion of the granules and the associated mass redistribution do not yield marked boundaries between dense granules and soft gels of the sort implied by Fig. 2.6. The matter has been investigated by applying material models derived from rheological and conductivity tests in the laboratory to simplified microstructural patterns using numerical calculation methods (finite element method, FEM), which has improved the understanding of the maturation process very significantly (Börgesson 1995). An important fact is that although the angle of internal friction of smectite clay is low, i.e. in the range of 8–15°, it is the reason for rather significant microstructural heterogeneity even after very long periods of time. Thus, density differences are never completely evened out and it has been found that the maximum and minimum densities in a clay element are in fact on the same order of magnitude as indicated in Fig. 2.6 and Table 2.1. However, the calculations show, as expected, that the density does not change abruptly at the interface between expanded granules and clay gel fillings in the external voids, i.e. there is a successive change in the spacing of flakes and stacks.

One example, with a mixture of 10% smectite clay and 90% sand, is shown in Fig. 2.7. The dry density of the 1-mm smectite grains was taken as 1.98 g/cm^3, and that of the mixture 1.56 g/cm^3 (2.00 g/cm^3 at complete water saturation). The system was assumed to be instantly saturated with electrolyte-poor water and the calculation was confined to describe the redistribution of mineral substance and porewater as a function of time. It was concluded from this study, which was made by use of the FEM code ABAQUS, that microstructural rearrangements in the maturation phase take place quickly, and that very considerable clay-gel density differences prevail in the larger part of the mixture after reaching equilibrium, i.e. 1.40–1.93 g/cm^3 (water-saturated state).

A very important fact is that the most narrow space in the contact region of adjacent sand grains will not be filled with clay and will hence contribute very significantly to the bulk hydraulic conductivity. At the boundary between this open space and the clay gel with density 1.40 g/cm^3 there is a zone with a clay-gel density that drops from this figure to 1.0 g/cm^3. Here, pressurized gas makes its way by displacing the soft gel and forming channels that become permanent. In repositories where hydrogen gas is produced by corrosion of iron canisters, reinforcement in concrete etc., it will penetrate clay-poor backfills at very low overpressure, i.e. a few tens or hundreds of kilopascals, in addition to the

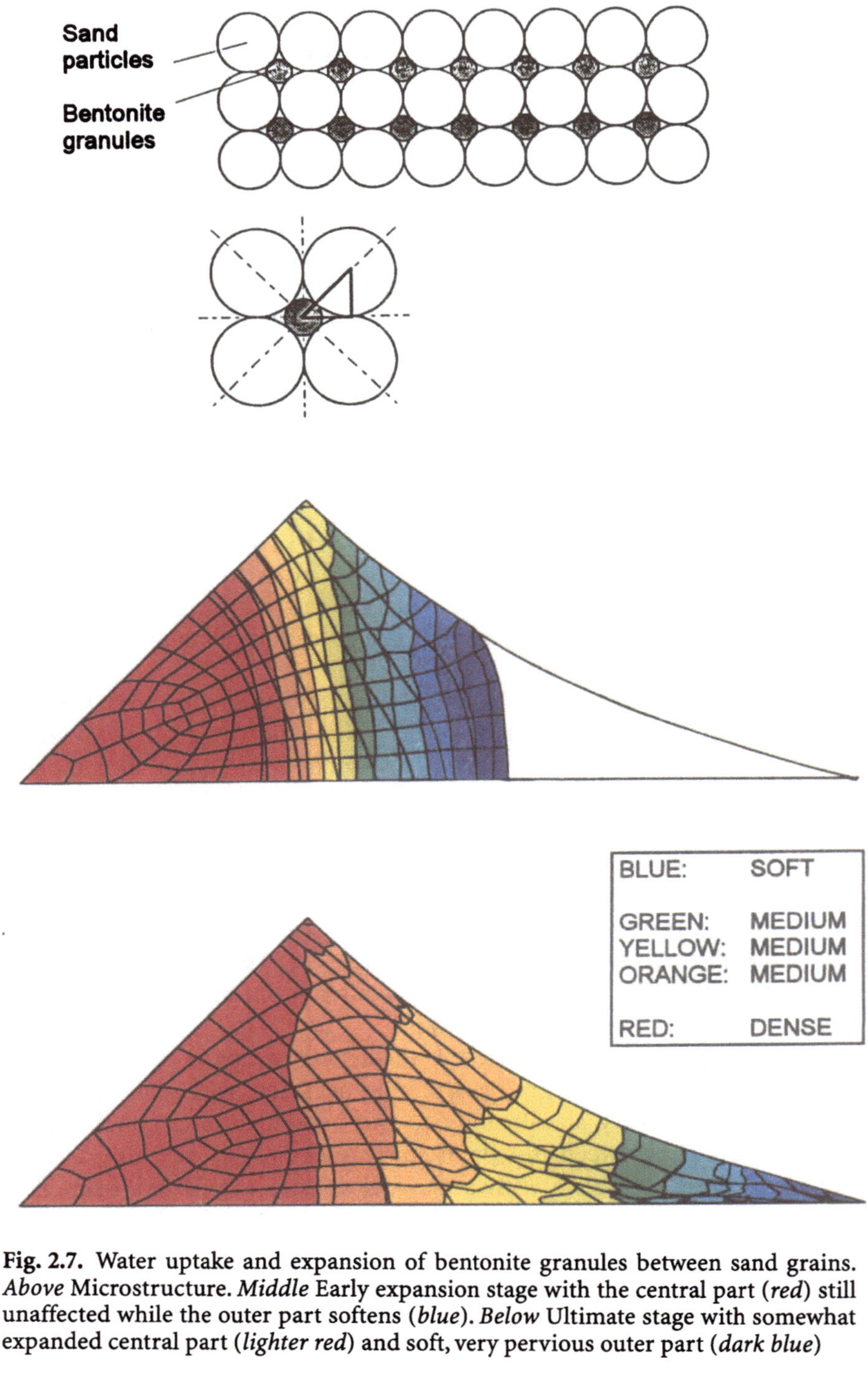

Fig. 2.7. Water uptake and expansion of bentonite granules between sand grains. *Above* Microstructure. *Middle* Early expansion stage with the central part (*red*) still unaffected while the outer part softens (*blue*). *Below* Ultimate stage with somewhat expanded central part (*lighter red*) and soft, very pervious outer part (*dark blue*)

prevailing piezometric pressure, while the "breakthrough" pressure is much higher in compacted bentonite, where the resistance to displacement of the clay matrix is much stronger (Pusch 1994b).

2.4
Physical Properties of Smectitic Clay

2.4.1
Hydraulic Conductivity

Figure 2.8 illustrates the dependence of density on the hydraulic conductivity of clays that consist of pure clay minerals (Pusch 1994a). The void ratio, i.e. the ratio of the pore volume and the volume of the solid substance, is a measure of the density, which is also given explicitly in the diagram. One finds that while kaolinite and hydrous mica (illite) have a rather low conductivity at high densities, montmorillonite is in fact less permeable even when the density is very low. Dense montmorillonite-

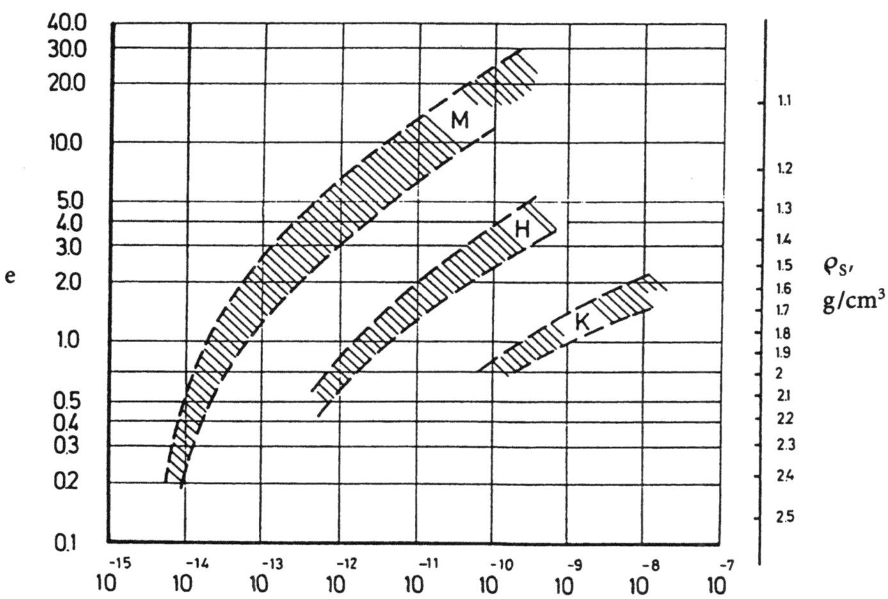

Fig. 2.8. Hydraulic conductivity of pure clays, the bands representing variations in hydraulic gradients and water chemistry. *M* Montmorillonite; *H* hydrous mica, *K* kandites (e.g. kaolinite); *e* void ratio; ρ_S density at complete water saturation

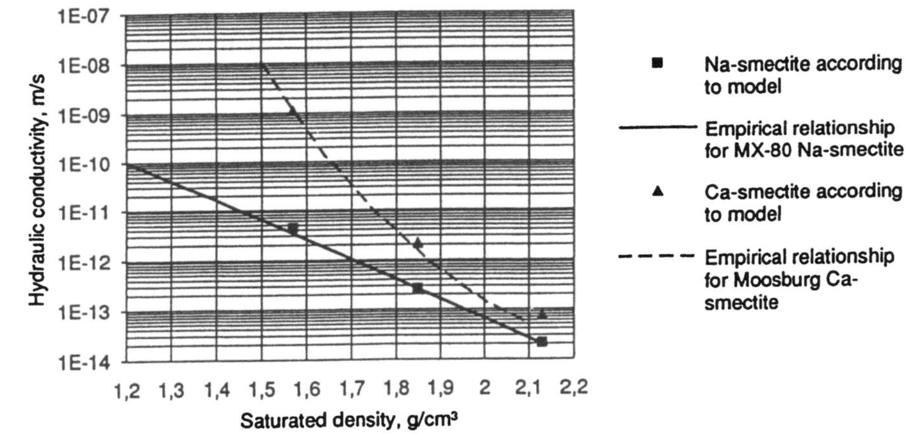

Fig. 2.9. Hydraulic conductivity of Na and Ca bentonite percolated by electrolyte-poor water. *Curves* represent empirical relationships; *triangles* and *squares* represent data calculated by applying microstructural models

rich clays are almost impermeable, which makes them suitable for embedding waste canisters and for constructing plugs in strategic positions in a repository. Since they are very slowly percolated by groundwater or not permeated at all, the transport of radionuclides through them primarily takes place in the form of diffusion.

The reason for the very low hydraulic conductivity of montmorillonite is that part of the porewater is in interlamellar positions and thereby largely immobile. As the diagram in Fig. 2.2 shows, the fraction of such water is over 80% when the dry density exceeds 1.2 g/cm³ (1.75 g/cm³ at water saturation). Applying the earlier mentioned geometrical data of the clay-gel fillings in the external voids and the number of hydrate layers that can be hosted in the interlamellar space of stacks in the granules and clay fillings, the permeable passages have been modelled and taken as a basis of calculations of the bulk hydraulic conductivity. As illustrated by Fig. 2.9, the agreement between predictions and experimentally determined conductivities is good. Figure 2.10 shows a device used for determining both the hydraulic conductivity and the swelling pressure.

2.4.2 Gas Conductivity

The generation of hydrogen by corrosion of steel canisters and canisters made of copper and steel, like the currently proposed Swedish type,

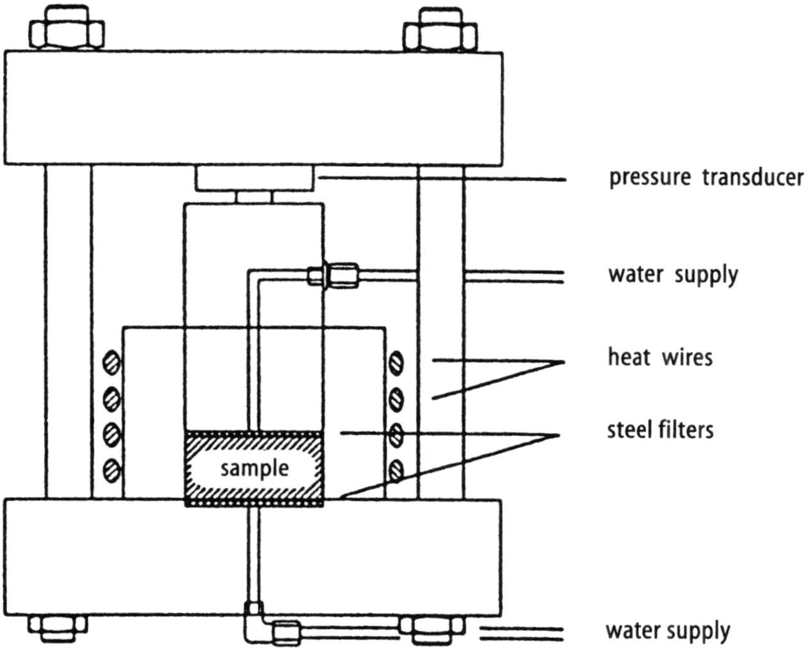

Fig. 2.10. Swelling-pressure oedometer. The sample is prepared by compacting the material in air-dry or wetted form in the cell. Maintaining a constant volume, water is let in and the resulting pressure recorded. After saturation, a hydraulic gradient is applied and the flow measured from which the hydraulic conductivity is evaluated

means that gas is produced and tends to expel water from the voids in the canister-embedding clay (Pusch 1994a, b). This requires that the gas pressure exceeds the prevailing piezometric pressure and what is commonly called the "capillary" pressure. The true gas-releasing mechanism does not have the nature of capillary retention, however; it is merely a matter of displacement of clay gel from external voids in clays with low or medium density, and of forcing densely arranged stacks of flakes apart in clays of very high density.

The very low shear strength of the clay gels when the bulk density is low, and of the clay fillings in the voids between ballast particles in bentonite/ballast mixtures of any density, means that the pressure required to make gas propagate is only a few tens or hundreds of kilopascals higher than the piezometric pressure. For very dense bentonite the corresponding "breakthrough" pressure is very high and approaches the swelling pressure, which is commonly several megapascals (Pusch 1994a).

2.4.3 Ion Diffusivity

The density of the clay, type of diffusing ion and ionic strength of the porewater solution are determinants of the diffusivity (Muurinen and Lehikoinen 1995). The diffusive anion-transport capacity is proportional to the ratio of the pore space of the gel fillings in the external voids and the total clay volume, while that of cations is – in principle – proportional to the ratio of the *total* void space and the total clay volume. This means that for montmorillonite clay with a bulk density at saturation of about 2 g/cm³, the diffusive transport capacity of anions like I- and Cl-, expressed in terms of the effective diffusion coefficient D_e, is only a few percent of that of important positively-charged radionuclides such as Cs^+ and Sr^{2+}.

Cation diffusion takes place in the form of pore and surface diffusion. The latter has the form of migration along basal planes of stacks of flakes and through the interlamellar space of the stacks. It follows from the significant difference between "internal" and "external" water space at varying bulk density that the cation diffusion capacity should be much higher than that of anions and this is also the case as indicated by Fig. 2.11.

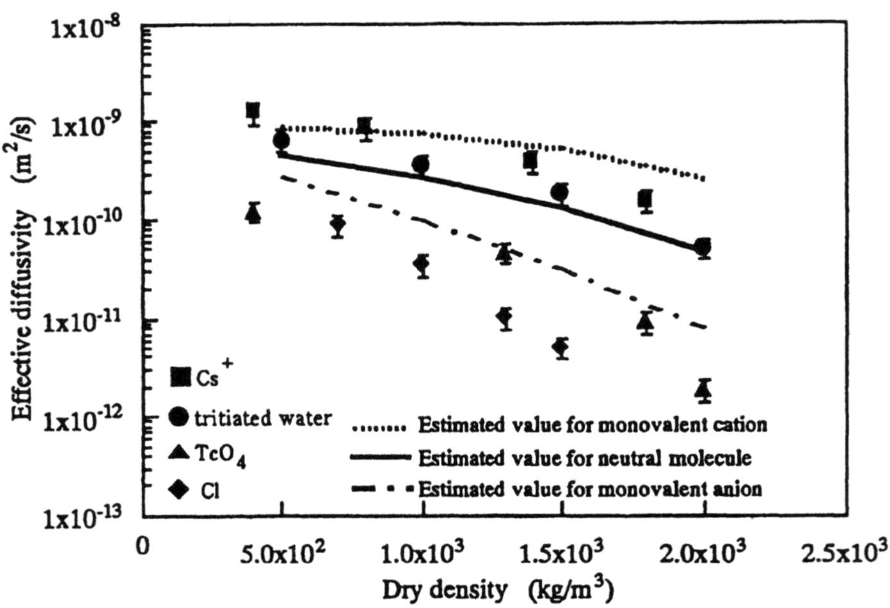

Fig. 2.11. Measured and calculated effective diffusivities for dense bentonite. (Kato et al. 1995)

One finds from the diagram in Fig. 2.11 that the order cation>neutral> anion can be identified both theoretically and experimentally. There are considerable differences in the diffusivity of different cations depending on their size, charge and position. Thus, only part of the sorbed cations are in the diffusive part of the electrical double-layer (Fig. 2.12) and hence mobile, while the rest are in the Stern layer and less mobile (Pusch et al. 1990).

Several investigations suggest that there are reasons to believe that surface diffusion – within and along the stacks of flakes – is a dominant migration mechanism (Muurinen and Lehikoinen 1995). It appears that the chemical conditions, in terms of the cation concentration and electroneutrality, control the diffusion rate, making it difficult to identify the importance of microstructural features. One condition of this sort is the concentration of competing cations, i.e. the ionic strength of the porewater, implying that an increasing strength decreases the sorption of radionuclides. Since the contribution of surface diffusion is determined by the amount of cations sorbed by ion exchange, its importance decreases when the electrolyte content increases.

Another physicochemical phenomenon is that exclusion of electrolytes arises when compaction induces overlap of electrical double-layers of adjacent stacks of flakes. Thus, cations are also excluded because they must remain with their anionic counterpart to maintain electrical neutrality (Muurinen and Lehikoinen 1995).

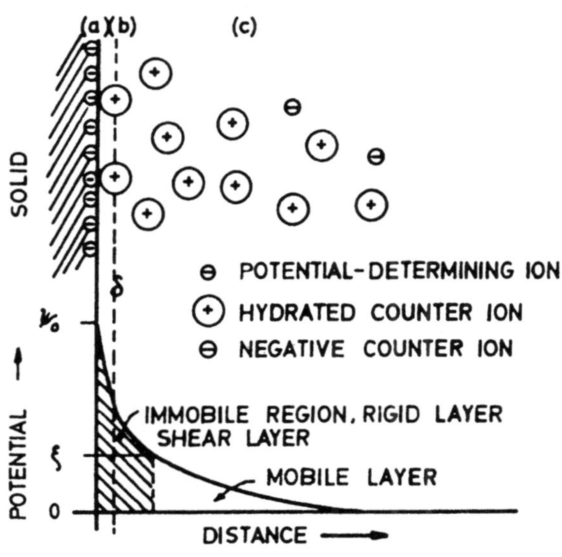

Fig. 2.12. Electrical double-layer at negatively charged mineral surfaces. **a** Surface charge. **b** Stern layer. **c** Diffuse layer of counter ions

Putting together the factors that are related to the microstructure and which affect the diffusion rate, one finds the following:

1. The higher the density, the more restriction to both cation and anion migration is caused by shrinking external voids with overlapping electrical double-layers.
2. The higher the density, the more effective and complete will be the occupation of interlamellar adsorption sites and bonding of interlamellar water and cations. This reduces the diffusion rate of hydrated cations but probably not of Na^+ and Li^+.
3. The higher the density, the more tortuous are the ion migration paths and the lower the diffusion rate of both cations and anions.
4. At very high densities the anion transport passages are hardly continuous and the anion diffusion rate is very low; since cations remain with their counterpart, the cation diffusion rate will also be very low with the possible exception of Na^+ and Li^+.

2.4.4 Swelling

Free swelling of smectite clay can be very significant since it implies that water is adsorbed to yield complete hydration of the interlamellar space and exfoliation of stacks of flakes to form a porous, cardhouse-like clay gel, in the same fashion as in the maturation of dense smectite clay (Pusch et al. 1990). Ultimately, i.e. when the negative porewater pressure induced by the hydration potential has evened out, the clay has undergone complete transformation to a very soft mass, several times larger in volume than the original clay.

If a sample of smectite clay is contained between the filters of a swelling pressure cell, like in the apparatus in Fig. 2.10 or in any space of constant volume and pervious surroundings like a deposition hole in the rock, free expansion cannot take place and a swelling pressure will instead be exerted on the confinement. This pressure can be very substantial and is valuable since it produces a tight contact with the rock in deposition holes preventing flow along the clay/rock contact. Figure 2.13 shows the successive build-up of the swelling pressure when air-dry, compacted bentonite powder is wetted in a swelling pressure oedometer to a density of saturation of 1.8 g/cm³. The initial pressure peak is caused by the hydration of the dense granules, which are in contact with each other and separated by voids that are essentially empty from the start. The subsequent slight pressure drop is due to sliding of sheared stacks into more stable positions, and exfoliation of stacks to form the clay fillings of external voids. The long-term trend to yield an ultimate maximum pressure is caused by completion of the hydration process.

If the fully water-saturated sample is allowed to expand in the oedometer by moving the upper piston upwards and filling the resulting open space with water, the clay expands in conjunction with absorbing water and will ultimately fill the space. If the wall friction is eliminated, the final microstructural constitution will be largely homogeneous in the entire expanded sample but the density will naturally decrease causing an increase in conductivity and a drop in swelling pressure (Börgesson 1990).

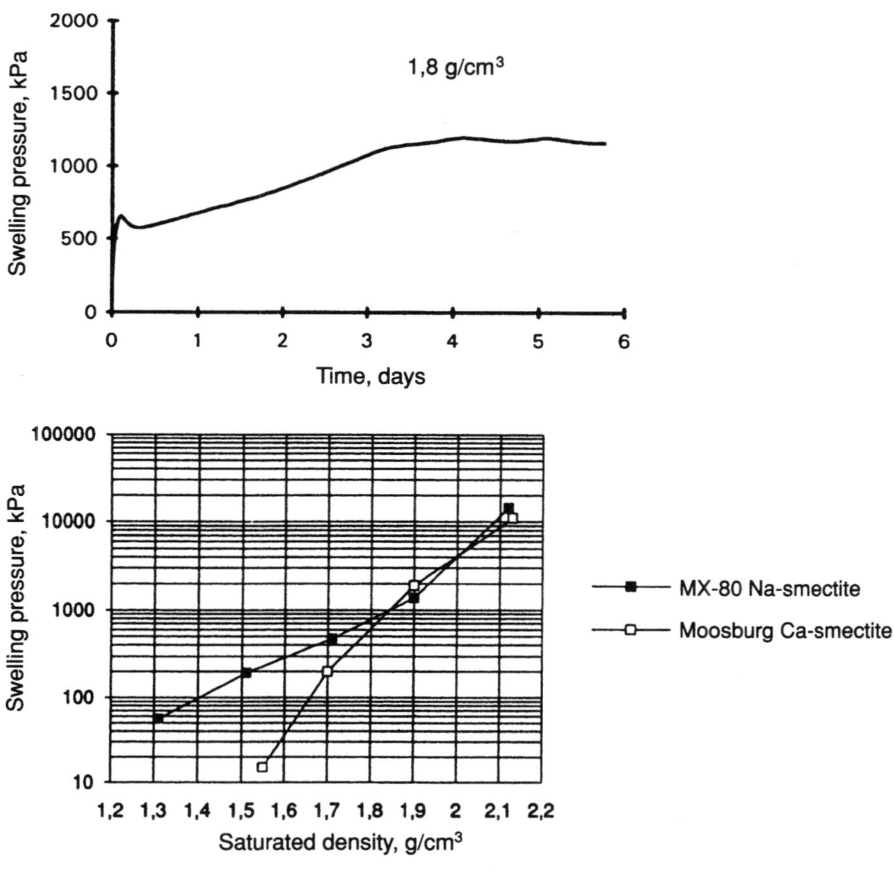

Fig. 2.13. Development of swelling pressure by hydration of initially air-dry bentonite powder. *Above* Buildup of swelling pressure under confined conditions. *Below* Fully developed swelling pressure of Na and Ca smectite

For Na montmorillonite, the maximum number of interlamellar hydrates is three, which means that a fully expanded state of the stacks is theoretically obtained for a bulk density at complete water saturation of around 1.8 g/cm^3, while the corresponding figure for Ca montmorillonite, which has up to two hydrates, is slightly lower (Pusch et al. 1990). This means that below these critical densities, the swelling pressure is entirely controlled by electrical double-layer repulsion.

Since the number of stack contacts, and therefore also of interacting electrical double-layers, is much larger in Na-montmorillonite than in montmorillonite saturated with bivalent cations, one would expect a stronger influence of cation concentration on the swelling pressure for the firstmentioned clay type. This means, in turn, that the swelling pressure for densities below the critical levels should be very low at any salinity for Ca montmorillonite and at high cation concentration for Na montmorillonite. At a low ionic strength it should increase from a very low value to an appreciable pressure for Na montmorillonite, when its density increases from a very soft state to the critical value. This is illustrated by the schematic relationships between density, cation concentration and swelling pressure indicated in Fig. 2.14 (Pusch et al. 1990).

The consequence of all this is that the physical properties of soft smectite clay are controlled by the chemical composition of the porewater. Thus, grouts injected in rock fractures for sealing them off may coagulate and become physically unstable if they are prepared with freshwater and the groundwater is saline, because their density is very low, i.e. 1.1–1.3 g/cm^3. On the other hand, if they are prepared with $CaCl_2$ solution and the groundwater is electrolyte-poor with Na as dominant cation, they form stable, low-permeable gels (Pusch et al. 1991).

2.5 Longevity

2.5.1 Physical Stability

By physical stability one means that the microstructural constitution of the clay is preserved, i.e. that channels or voids that significantly change the bulk physical properties are not formed, and that the clay does not shrink and lose contact with the surrounding rock or waste canisters. In soft smectite clay, coagulation due to changes in porewater chemistry can take place as discussed above, and piping and erosion by penetrating water or gas may occur if the local hydraulic gradient or the gas pressure

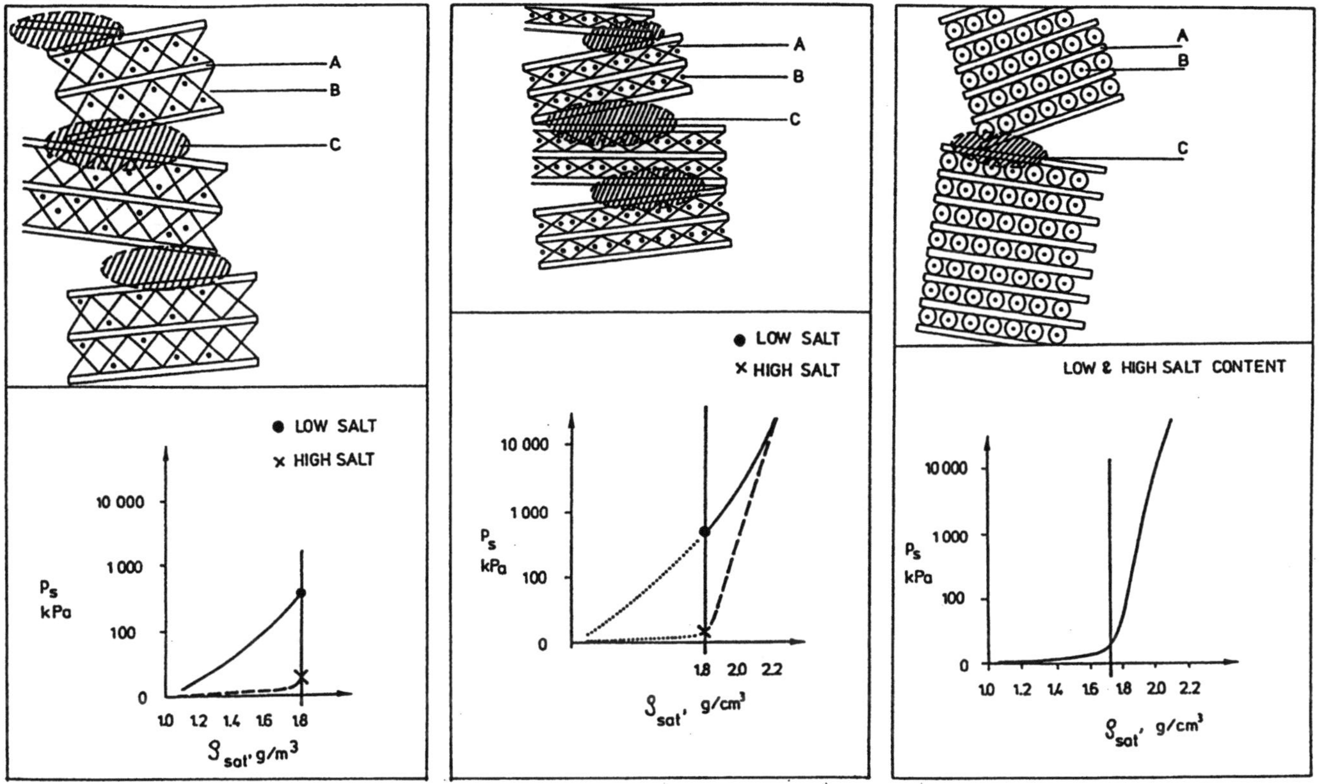

Fig. 2.14. Schematic pictures of stack assemblages and influence of bulk density at water saturation and salinity for Na and Ca montmorillonite clay. **A** Flake; **B** interlamellar space; **C** stack contact region with interacting electrical double-layers. *Left* and *central* pictures represent Na clay; *right* Ca clay. (Pusch et al. 1990)

are critically high. In dense bentonite, microstructural discontinuities or shrinkage can take place by drying, mineral alteration or gas penetration. Drying may temporarily occur close to waste canisters in the period when radioactive decay produces significant heat, and it may lead to a permanent drop in expandability of the stacks of flakes because of cementation effects. The most important process is mineral alteration, which is controlled by the chemical stability of smectite.

2.5.2 Chemical Stability

The major source of information on the chemical stability of smectites has been geological data from deep drillings and analyses of clays that have undergone hydrothermal treatment under controlled conditions in the laboratory. The most important degrading process is concluded to be conversion of the smectite to illite, or more correctly hydrous mica, but cementation by precipitation of silica/aluminium compounds formed from dissolved smectite in the hottest part of canister-embedding clay may also be important (Pusch 1993).

Most geological data originate from the Gulf area and from recent investigations of North Sea sediments. For the Gulf it has been found that smectite-to-illite conversion has usually proceeded from about 80% smectite (S) and 20% illite (I) to 20% S over the temperature range 50–100°C, resulting in regularly stacked mixed-layer minerals (i.e. ISIS, etc.). The overall reaction has been claimed to be:

$$S + K^{+} + Al^{3+} = I + Si^{4+}. \tag{1}$$

Despite several years of intense research, the exact conversion process is not known. One possibility is that silicons in tetrahedral positions of the smectite are replaced by aluminium emanating from octahedral positions of the smectite lattice in conjunction with occupation of the later sites, e.g. by magnesium, and that the released silica forms quartz, cristobalite or amorphous silica. Such "beidellitisation" requires that magnesium or other cations are available and they may be offered by groundwater electrolytes or dissolved accessory minerals. Another possibility is that smectite is dissolved, giving off silica and aluminium, which are used up in the precipitation of illite to an extent and at a rate that is controlled by the presence of dissolved potassium. This idea, i.e. that of neoformation of illite, has been supported by a number of researchers, e.g. Nadeau et al. (1985), who showed that the concept of mixed-layer I/S is questionable. Still, the question whether smectite-to-illite conversion takes place through mixed-layer formation or by dissolution/neoformation remains open (Pusch 1993).

A large number of experimental hydrothermal investigations were performed in the 70s and they have all indicated that smectite is converted to illite following Arrhenius' law. This was also the conclusion from modelling of such conversions in nature, like North American and Japanese bentonites with relatively well known temperature histories, from which the activation energy for conversion was found to be in the order of 27 kcal/mol. A general theory, based on Arrhenius' law and introducing the K^+/Na^+ ratio as an essential parameter, was proposed by Pytte (1982), and it has been found to yield values in fair agreement with actual smectite-to-illite conversion rates both in nature and in laboratory experiments. The conversion rate has the following form:

$$-dS/dt = Ae^{-U/RT} f(K^+/Na^+)\, g(S), \qquad (2)$$

where t is time and S the mole fraction of smectite in I/S assemblages, A is a constant and U the activation energy, R is the gas constant, T is the temperature in degrees absolute, K^+ and Na^+ the concentrations of dissolved potassium and sodium, and f and g denote functions of K/Na and the smectite content.

Using Pytte's conversion law with appropriate coefficients and the activation energy 27 kcal/mol, one gets the diagram in Figure 2.15 for the

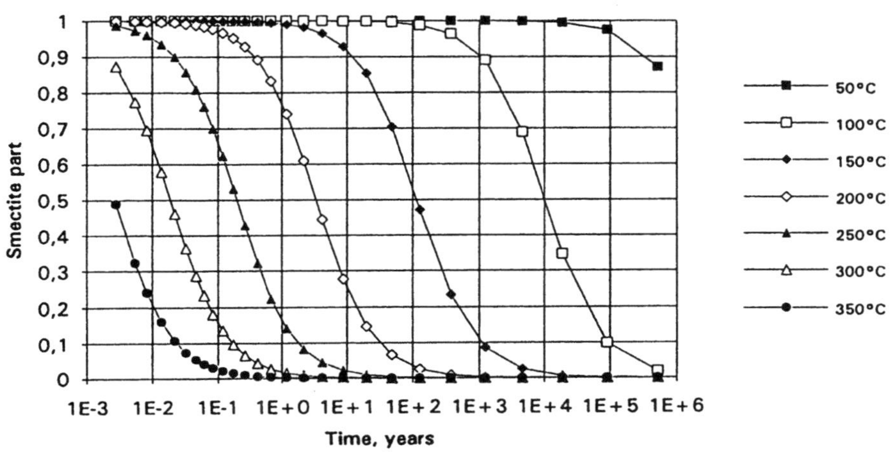

Fig. 2.15. Rate of conversion of smectite to illite according to Pytte's law for the K^+ concentration 0.01 mol/l, and the activation energy 27 kcal/mol. The initial smectite content, which is taken as 100% (smectite part = 1), drops successively with time and temperature

temperature- and time-dependent alteration of smectite to illite at a K^+ concentration of 0.01 mol/l. One finds from this diagram that at 100 °C, for example, only a few percent of the smectite will be converted to illite in a few thousand years, while 50 % is expected to be altered after a few hundreds of thousands of years. The assumed potassium concentration is commonly found at 500 m depth in crystalline rock, but higher figures may prevail in areas where seawater infiltrates. Under such circumstances, and at a temperature of 200 °C, the conversion is much quicker, and the majority of the smectite may in fact be altered in a few tens of years.

Although Pytte's law appears to be practically applicable, it is clear that the choice of parameter values, in particular the activation energy, is critical and makes predictions rather uncertain.

Investigation of the mineral composition of the immediate surroundings of natural bentonite beds that have been exposed to heat has shown that they are characterized by excess silica. This has been taken as verification of Eq. (1) but may equally well be the result of dissolution of accessory minerals, primarily feldspars, when illite is neoformed. Irrespective of the origin of the silica, it is inferred to serve as a cementation agent. Hence, it may weld stacks of flakes together and reduce or eliminate expansion of the remaining, unaltered smectite.

In deposition holes with highly radioactive waste, heat is produced in conjunction with the radioactive decay and a temperature gradient formed over the canister-embedding smectite clay, which causes silica cementation. Thus, silica released from smectite and accessory minerals near the hot canister, and not used up in the illitisation process, will diffuse towards the outer, cold part of the canister embedment and precipitate there. This process may be important for several hundred years at temperature gradients exceeding 1–2 °C/cm (Pusch 1994a).

2.6
Performance Under Repository Conditions

2.6.1
Groundwater Percolation Through Canister Embedment

After emplacement of the waste canisters and their bentonite embedment, water entering the deposition holes is absorbed by the strongly hydrophilic clay, which ultimately becomes completely water-saturated. The temperature gradient tends to drive porewater outwards from the vicinity of the hot canisters towards the colder rock, which initially causes some

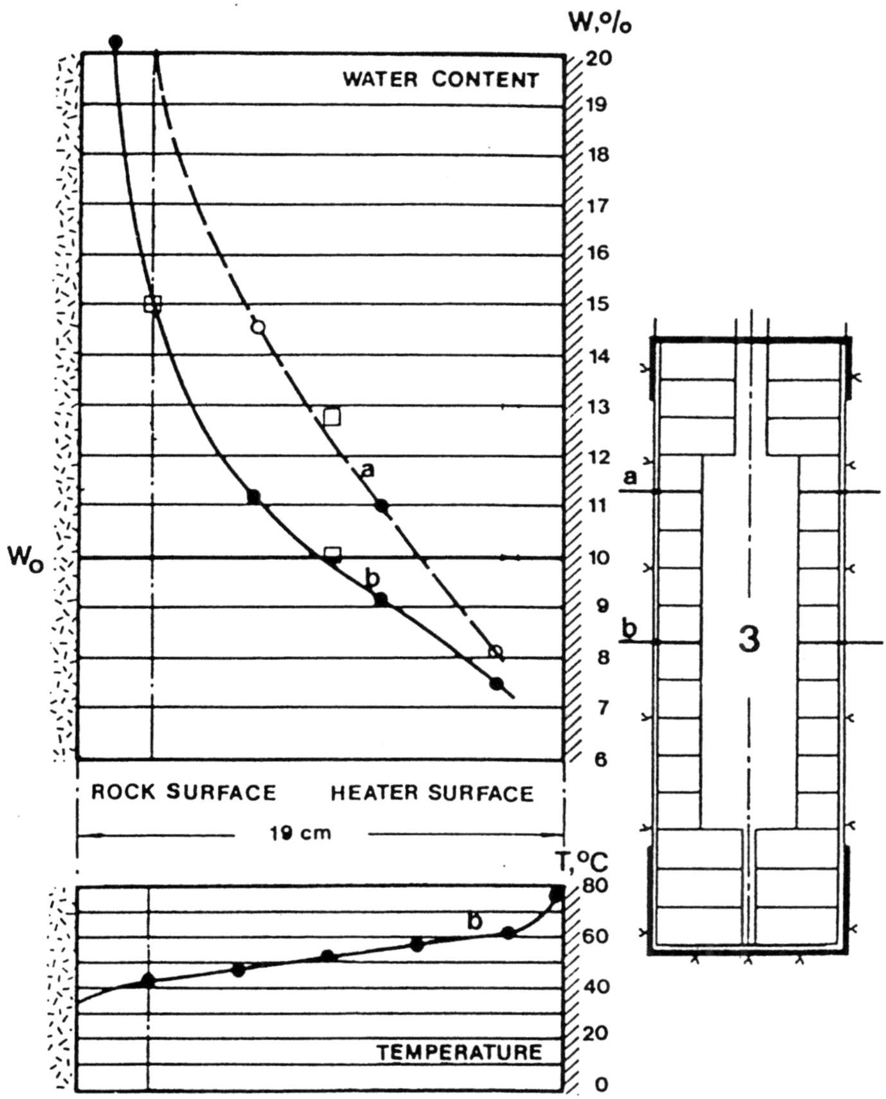

Fig. 2.16. Example of redistribution of the original water content and some water uptake early in the wetting process. Experiment at Stripa (central Sweden) with an electrical heater simulating a heat-producing canister (Pusch 1994a). The initial water content by weight (w_0) was 10%. Close to the heater this dropped to less than 7%, while transfer of moisture and uptake from the rock gave complete water saturation at the clay/rock contact (w = 20%)

drying of the bentonite close to the canisters and cyclic migration of water vapour from the hottest part to the inward advancing wetting front (Fig. 2.16). This leads to accumulation of dissolved salt and precipitation of certain carbonates and sulphates in the inner part of the bentonite envelope, which is assumed to cause cementation and partial loss of the expandability and self-sealing ability of the clay.

In the wetting phase, which lasts for a period of a few years up to a few tens of years, depending on the water-bearing capacity of the rock surrounding the deposition holes, no water flows through the canister embedment. At the end of the wetting process and thereafter, such flow can take place, but the flux will be extremely small because of the very low hydraulic conductivity of the bentonite and of the very low hydraulic gradients that prevail, i.e. $10^{-3}-10^{-2}$, after development of stable piezometric conditions. The absence of significant percolation by flowing water of the canister embedment means that radionuclide transport will take place almost entirely by diffusion.

2.6.2
Ion Diffusion

Radionuclides are released if and when water enters the canisters, which can take place through defects of the canisters like leaking welds or corrosion pathways. The solubility of burnt-out fuel, which is the high-level radioactive waste that will be enclosed in Swedish canisters, is very low, and the decay rate of the radioactivity sufficiently high to prevent breakthrough of certain radionuclides. Other species, like iodine, caesium and strontium, will not undergo substantial decay, and represent the actual risk to the biosphere. However, their transport to shallow depth through the rock can be strongly retarded by locating the deposition holes and tunnels such that, effectively, water-bearing rock features are avoided, and by applying seals in strategic positions in the tunnels and shafts. Migration of radionuclides is also effectively counteracted by adsorption on mineral surfaces in the rock matrix. Furthermore, the richness in fractures, and higher porosity of shallow rock than of the bedrock at repository depth, mean that dispersion takes place causing a very significant drop in concentration of radioactive matter.

2.6.3
Gas Formation

Canisters made of steel or containing steel give off hydrogen gas when they undergo aerobic and anaerobic corrosion. If the gas cannot be released through the clay embedment, gas bubbles would be formed at the clay/canister interface and cause displacement and heterogeneity of the clay. This matter has been investigated experimentally and a conceptual model developed for gas transport that is based on the general microstructural model referred to earlier in this chapter (Pusch 1994b).

Gas makes its way through the major pathways formed by interconnected "external" voids in the clay matrix. In smectite clays of very high density, gas penetration implies that dense stacks of flakes have to be forced apart, which requires that the pressure is on the same order of magnitude as the swelling pressure, i.e. several megapascals. The penetration has the form of rapid breakthrough when the critical pressure, i.e. the sum of the breakthrough and piezometric pressures, has been reached. In less dense smectite clays, gas makes its way by displacing and deforming the clay-gel fillings of external voids, which means that penetration takes place at considerably lower pressure.

2.6.4 Radiation

Even very thick metal canisters do not cut off gamma radiation from highlevel waste, and the clay embedment will therefore be exposed to this form of radiation (Pusch 1994a). Soon after the waste is placed in the deposition holes, this leads to radiolysis of the clay porewater, forming oxygen and hydrogen, and this has some minor effect on the corrosion of the canisters. The influence of gamma radiation on smectite minerals has the form of degradation of the crystal lattice, but it is not significant even considering the total radiation dose that the clay will be exposed to. Experiments have shown that the major effect is a reduction in particle size, which is not hazardous to the function of the clay.

Alpha radiation is negligible, with the exception of the effect that is caused by radionuclides that migrate by surface diffusion (Pusch 1994a). Thus, the very small distance of such radionuclides from the clay mineral lattices causes severe degradation and transformation to an amorphous state with complete loss of the expandability and hydration ability that is characteristic of smectite.

2.6.5 Microbial Processes

It is well known that microbes can carry radioactive matter and since they are present in bentonite and also in the rock, the matter of mobility of bacteria has been considered in performance analyses. These analyses have shown that bacteria cannot move in very dense smectite clay, like the canister embedment, because of the very small pore size and limited connectivity of the external voids, and because of the attractive forces that are developed between the negatively charged bacteria and positively charged parts of stacks of flakes, and because of hydrogen and van der Waals' bonds between closely-located bacteria and minerals. They hence turn into a dormant state and degrade and die by attack of enzymes (Pusch 1994a).

In smectite clay with a dry density lower than about 1.0 g/cm^3 (1.8 g/cm^3 at complete water saturation), bacteria may migrate and consume other organic matter, yielding organic colloids that can be transported through the clay, and further to the biosphere through rock fractures.

2.6.6
Rock Shearing

The fact that radionuclides cannot be released from the waste canisters unless their interior is exposed to water means that the matter of tectonic impact on the integrity of the canisters is of major importance. Thus, shearing of the canisters by displacement of rock along joints that intersect the deposition holes will cause shear failure and fracturing if the movement is sufficiently large (Fig. 2.17). Statistical analyses show that a deposition hole with 1.6 m diameter and 8 m depth in ordinary granite rock will be intersected by three to six discontinuities that can be mechanically activated and that one or two of them are flat-lying or slightly inclined, hence representing possibly critical conditions (Pusch 1995).

The risk of such displacements depends on the structural constitution of the rock and of the rock stress conditions and it can be assessed by performing numerical analyses (Pusch 1994a, 1995). Assuming the rock to form a fractal-like system of weaknesses with a shear resistance that is related to their size and nature (Fig. 2.18), current investigations indicate that severe earthquakes yielding almost instantaneous shearing along major weaknesses, i.e. shear zones of the type marked 1 and 2 in Fig. 2.18, produce shearing of discontinuities that intersect deposition holes (4 in Fig. 2.18) by less than 1 mm. This is about 1/100 of what a waste copper canister of Swedish type can sustain when it is embedded in a very dense and hence stiff surrounding clay. A large number of such displacements is expected to cause accumulated shear strain, but it is still estimated that critical conditions will not be caused even in a 100000 year perspective in Swedish bedrock. Naturally, location of a repository in or close to a major weakness or intersection by many fracture zones of type 3 in Fig. 2.18 is not suitable. It is also clear that there are many

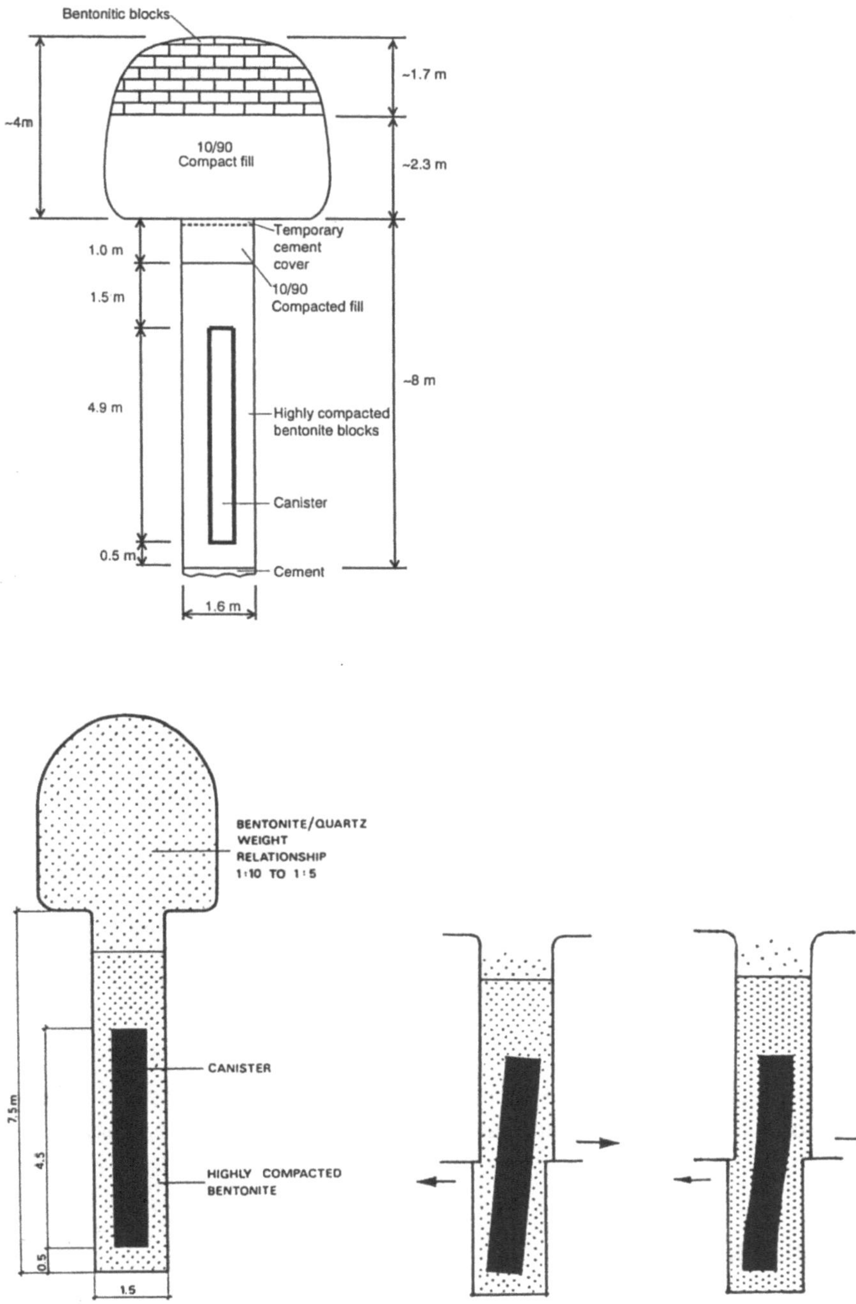

Fig. 2.17. Waste canister embedded by dense bentonite according to the Swedish KBS-3 concept. *Above* Schematic section of disposition holes. *Below* Mode of shearing that can cause breakage of waste canisters. (Pusch 1994a)

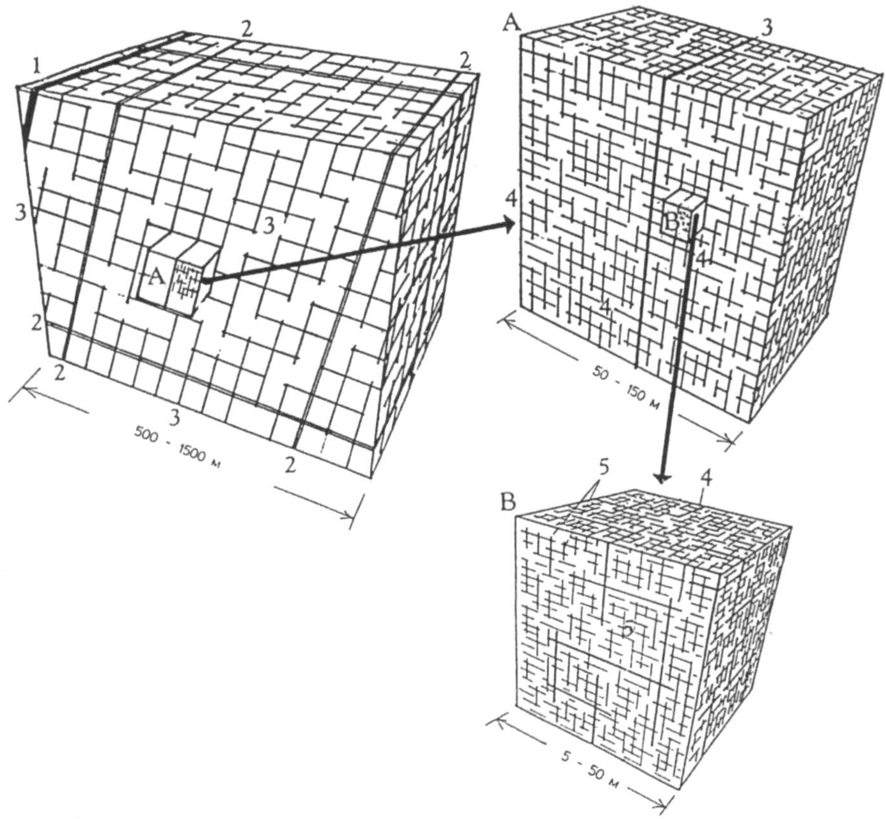

Fig. 2.18. Schematic view of orthogonally-structured granitic bedrock with the hierarchy of weaknesses that is commonly observed. *1* and *2* Major discontinuities like shear and crushed zones; *3* minor fracture zones; *4* discrete weaknesses that will appear in all deposition holes; *5* minor discrete weaknesses that can be activated by blasting or stress changes induced by excavation of repository. (Pusch 1994a, 1995)

areas in the world where the rock stress characteristics represent critical conditions, like central Canada and the rim areas of most continents.

2.7 Concluding Remarks

In conclusion, dense smectite clay in the form of precompacted bentonite blocks serves as a very effective and long-lasting isolation of hazardous radioactive waste. A number of independent reviews and current safety analyses performed by SKB, Swedish Nuclear Fuel and Waste Management Co., and Swedish authorities, show that repository concepts of the Swedish KBS3 type that make use of bentonite as a major isolating component, offer acceptable safety, even in a very long time perspective. The critical parameter is the temperature and the temperature gradient. As long as the former is lower than about 100°C and the latter less than about 1°C/cm distance from the canister surface, as in the KBS3 concept, the performance is concluded to be acceptable, but significantly higher values may imply poor performance of the clay.

References

Börgesson L (1990) Laboratory testing and computer simulation of clay barrier behaviour. Proc 9th Int Clay Conf, 1989, Strasbourg 1990, pp 117–126

Börgesson L (1995) Swelling and homogenisation of bentonite granules in buffer and backfill. SKB Arbetsrapport AR, SKB, Stockholm

Kato H, Muroi M, Yamada N, Ishida H, Sato H (1995) Estimation of effective diffusivity in compacted bentonite. In: Murakami T, Ewing RC (eds) Scientific basis for nuclear waste management XVIII. Mat Res Soc Symp Proc, vol 353. MRS, Pittsburgh, pp 277–284

Muurinen A, Lehikoinen U (1995) Evaluation of phenomena affecting diffusion of cations in compacted bentonite. Report YJT-95-05, Voimayhtiöiden Ydinjätetoimikunta, Helsinki

Nadeau PH, Wilson MJ, McHardy WJ, Tait JM (1985) The conversion of smectite to illite during diagenesis: evidence from some illitic clays from bentonites and sandstones. Mineral Mag 49:393–400

Pusch R (1993) Evolution of models for conversion of smectite to non-expandable minerals. SKB Technical Report TR 93-33, SKB, Stockholm

Pusch R (1994a) Wase disposal in rock. Developments in geotechnical engineering 76. Elsevier, Amsterdam

Pusch R (1994b) Gas transport through smectitic clay and crystalline rock. SKB Arbetsrapport AR 94-61, SKB, Stockholm

Pusch R (1995) Rock mechanics on a geological base. Developments in geotechnical engineering 77. Elsevier, Amsterdam

Pusch R, Karnland O, Hökmark H (1990) GMM A general microstructural model for qualitative and quantitative studies of smectite clays. SKB Technical Report TR 90-43, SKB, Stockholm

Pusch R, Karnland O, Hökmark H, Sandén T, Börgesson L (1991) Final report of the rock sealing project – sealing properties and longevity of smectitic clay grouts. Stripa Project, Technical Report TR 91-30, SKB, Stockholm.

Pytte AM (1982) The kinetics of the smectite to illite reaction in contact metamorphic shales. Thesis, Dartmouth College, NH Dartmouth, New Hampshire

Chapter 3

Roles of Clay and Oxide Minerals in Soils[1]

J.B. Dixon

3.1 Introduction

Soils are complex mixtures of minerals, organic matter, and water. The most active mineral constituent of soils is clay. This is because of the high surface area of the fine particles and because of the inherent structural features of certain clay minerals. Most soils contain only a few percent organic matter but a much larger amount of clay. The lack of any orderly atomic arrangement in soil organic matter prevents discussing it in an analogous way to minerals, which are predominantly crystalline constituents. Thus, this discussion will be largely devoted to the clay minerals, i.e. the minerals in the fraction 2 micrometers (µm) and less in equivalent spherical diameter.

Soils are used for many purposes, e.g. for the growth of plants, for the support of buildings, roads and airports, and for the containment of lakes. Thus, they require many varied properties to serve the purposes to which they are applied.

3.2 Color

The color of soil is largely caused by the presence of certain minerals (Schwertmann and Taylor 1989; Bigham and Ciolkosz 1993). The Fe oxide haematite gives the soil a red color. The presence of goethite as the only Fe oxide gives the soil a yellow brown color. Manganese oxides make certain

---

[1] Reprinted (with modifications) from Applied Clay Science, 5, edited by H.H. Muray, J.B. Dixon, Roles of Clays in Soils, pages 489–503, 1991, with kind permission from Elsevier Science - NL, Sara Burgerhartstraat 25, 1055 KV Amsterdam, The Netherlands.

soil features black but they are seldom abundant enough or well enough dispersed to influence the color of the whole soil. Nodules and some particle surfaces are often black due to the presence of Mn oxides. Understanding soil color requires knowledge of soil constituents. For example, a soil that is red due to the presence of haematite may also contain a large part of its Fe oxide as goethite. Likewise, black soil concretions may contain more Fe than Mn oxides due to the dominance of black over other colors, e.g. black nodules in the Decatur soil of Alabama (Uzochukwu and Dixon 1986). Poorly drained soil such as those in the coastal region south of Houston, Texas contain black dendritic coatings on the surfaces of aggregates visible under a 10 × lens. Apparently, Mn oxides precipitated there due to seasonal oxidation. Organic matter contributes to the general darkening of soils too, particularly in the surface layer and less frequently in layers below the surface (e.g. smectitic clay subsoils).

3.3 Potassium in Soil Minerals

The native K supply in soils that gradually becomes available to plants is primarily from micas (or illites); some are clay size, others are larger. In the younger soils such as the loessial soils of the midwestern USA (Mollisols and Alfisols, commonly), native K is usually high and the release rate is important to plant nutrition. In older soils such as the acid highly weathered soils (primarily Ultisols) of the southeastern USA the mica content is low, more resistant to weathering, and much greater reliance on applied K fertilizer is necessary for economic plant production. Since K is an essential element that is required by plants in large quantities, fertilization with K is sometimes necessary even when native supplies are high, depending on the crop being grown. The release of K from soil clay and the coarser fractions has an great literature reflecting its importance to crop production and world food supply. Hydroxyl orientation, di- versus trioctahedral form of mica, particle size, oxidation-reduction conditions, and layer charge are some of the factors that have been credited with influencing the adsorption and release of K from micas and their weathering products. The subject of K in soil minerals was reviewed in four papers that were introduced by Robert (1987).

3.4 Cation Exchange Capacity

The importance of layer silicate clays to the properties of soils is best illustrated by their cation exchange capacity. The presence of permanent negative sites in certain clay minerals accounts for the ability of many soils to hold cations against leaching, yet in an exchangeable form that supplies nutrients to plants. Also, these exchangeable ions have variable hydration spheres that influence the shrinking and swelling of clay soils as the water content changes.

Smectite, vermiculite, and zeolite are the groups of minerals that have the highest cation exchange capacity in soils. Their charge is predominantly from ionic substitution in structural positions. Smectites are the most common group and, owing to their small particle size and high charge, are the most chemically and physically active minerals in soils. Smectite is a typical clay mineral, rarely occurring as coarse particles (i. e. silt or sand), and ions are usually freely exchanged although highly charged smectites with more ion selectivity have been reported in some Vertisols (Dixon 1994). Vermiculite and zeolite minerals have very high negative charge and occur most often in coarse clay and silt fractions. Vermiculite is recognized for its selective retention of certain large cations and its limited shrink-swell range (see later). Zeolite particles are highly charged and have ion selectivity controlled by the size of tunnels in their structures that restrict the passage of ions and molecules (Ming and Mumpton 1989).

3.5 Shrinking and Swelling of Soils

Many soils, particularly those that contain abundant clay, shrink and form visible cracks as they dry. The shrinking of soils is credited with forming pressure faces that contribute to the boundary of soil structural units (aggregates). The development of structural units in soils is important to their aeration and other physical properties that are troublesome in unstructured or puddled clay soils. On the other hand, shrinking and swelling of soils causes billions of dollars in damage to roads and buildings in the USA every year (Jones and Jones 1987). Smectite clay is largely responsible for the shrinking and swelling damage of most soils. This group of minerals, due to their fine particle size, high surface area and variable interlayer spacing, is extremely active physically and chemically.

Because of the special properties of soils that contain large amounts of smectite clay, a separate category has been established in soil taxonomy, Vertisols. Vertisols are recognized for their high coefficient of linear extensibility, gilgai microrelief, variable depth and other unique properties throughout many parts of the world (Dudal and Eswaran 1988). In the USA, they are most extensive in Texas but they occur in many other states in localized areas.

The other clay minerals that have large amounts of associated water and potential shrinkage are allophane and imogolite. These minerals are extremely fine grained (Wada 1989). The particles are so small that they are sometimes considered to be noncrystalline. They are most abundant in soils formed from volcanic ash, thus they occur primarily in certain parts of the world where volcanic parent materials prevail. Alaska, Hawaii, Oregon, Washington in the USA and Japan, New Zealand and Iceland (Arnalds et al. 1995) contain areas of significant occurrence of these soils now called Andisols, a soil order designed for them (Leamy et al. 1980; Soil Survey Staff 1990). Andisols occur in regions of uniformly moist conditions, therefore they do not frequently shrink. When they do dry out their volume change is great and it is not as reversible as it is in Vertisols (Warkentin and Maeda 1980).

3.6 Practical Problems with Smectitic Clays

Too much smectite clay in soils leads to several problems of agronomic significance (Dudal and Eswaran 1988). Smectitic soils can be tilled only in a narrow moisture range thus making timing critical. Moist Vertisols are cohesive and difficult to plough (Randolph and Reed 1938). It takes four times as much power to plough a Houston clay (Vertisol) as a Norfolk sand, and it takes twice as much power to plough the Houston clay as a Sumpter clay that contains a similar clay mineral suite, but the Sumpter soil contains about 66 % calcium carbonate.

Soils that contain large amounts of smectite have numerous slickensides, attributed to their history of shrinking and swelling (Fig. 3.1). These planes of weakness portend sidewall failure when they are excavated. The planes of weakness in Vertisols are also passageways for water around soil structural units. Thus, the ion movement through these clayey soils is localized via the larger pores and may lead to loss of nutrients from the plant root zone and rapid transport of contaminants (Ritchie et al. 1972; Southard and Graham 1992). Other well-structured clay soils exhibit nonuniform movement of water (bypass flow), thus contributing to the complexity of their behavior as a medium for plant growth and for waste utilization or disposal (Booltink and Bouma 1991).

3.7 Selective Ion Retention

Ion retention by clay minerals is achieved by several mechanisms and the energy by which they are held also is varied. The typically free exchange

Fig. 3.1. Slickenside in a Vertisol: a product of smectite shrinking and swelling

of cations was discussed earlier. Yet there are several examples of cations being held so strongly that they are not exchanged by another ion of the same charge (Sawhney 1972). Vermiculite selectively holds K, NH_4, Cs, and Rb and there is mounting evidence that some smectites do also. Obviously ion fixation may be helpful or detrimental depending on the use to which the soil is applied. The adsorption of radioactive Cs by clay is of great environmental significance in a nuclear age. On the contrary, the fixation of NH_4 in rice paddy soils can reduce economic production of grain (Chen et al. 1989).

Manganese oxides selectively adsorb Cu, Co, Ni, Ba, Pb and Zn (Golden et al. 1986; McKenzie 1989). The mechanism of retention probably is different between Co and the other ions. The adsorption of Co is believed to be a special case where it is oxidized to Co^{3+} and enters structural sites and replaces Mn (McKenzie 1989).

Anions are also selectively held by some clay minerals in soils. Phosphate adsorption is widely recognized to be greater in soils that contain more iron oxides. By a careful x-ray diffraction analysis procedure, Jones

(1981) was able to establish a close correlation between phosphate adsorption and the surface area index of the goethite present.

3.8 Soil Minerals Versus Model Minerals

There are important differences between minerals formed in soils and the model minerals that occur in relatively pure deposits or that can be synthesized. This diversity in mineral properties is illustrated by the development of hydroxy-Al interlayers in vermiculite (Fig. 3.2) and by the substitution of Al for Fe in the structural positions of goethite and haematite. In 1949, Pearson and Ensminger found a 14-Å mineral in an Alabama soil that had properties different from vermiculite, smectite or chlorite, and MacEwan (1950) reported a 14-Å reflection that became diffuse on heating at 540 °C. Many investigations indicate that soil vermiculite and some smectite are altered to a 14-Å phase that contains positively charged hydroxyl-Al groups that restrict closure to 10 Å and reduce cation exchange capacity and surface area. Also, the surface properties of these minerals are altered. Current concepts are that the hydroxy-Al interlayers have a OH/Al ratio of 2.7 for the adsorbed Al component (Barnhisel and Bertsch 1989). Since that early work, hydroxy interlayered vermiculite has been recognized in soils from many parts of the world. It is unique in that it is most abundant in surface soil horizons. Also, it is typically concentrated in the coarse clay fraction (2 to 0.2 µm). X-ray diffraction, chemical analyses and many other methods have been applied to this mineral but the exact interlayer structure is still hypothetical (Fig. 3.2). While the investigation of interlayering in soil minerals has continued, there has been a parallel interest in similar so called pillared compounds, prepared by thermally treating hydroxy interlayered clays for use as catalysts (Zielke et al. 1989).

The presence of Al for Fe substitution in goethite of soils of Australia was shown by Norrish and Taylor (1961). It is inversely correlated with crystal size (Fig. 3.3). Once again, the presence of Al in the soil solution has modified the properties of a mineral in soils. Haematite has Al in its structure but only to about half the extent that it is in goethite (Schwertmann and Taylor 1989). The presence of Al in these iron oxides reduces their solubility. Thus, it appears that Al contributes to the stability of these oxides in a very finely divided state in soils. Goethite and haematite occur as larger crystals in some soil nodules and frequently in pure forms, yet their most common and most important occurrence in soils is in the fine clay fraction, i. e. less than 0.2 µm.

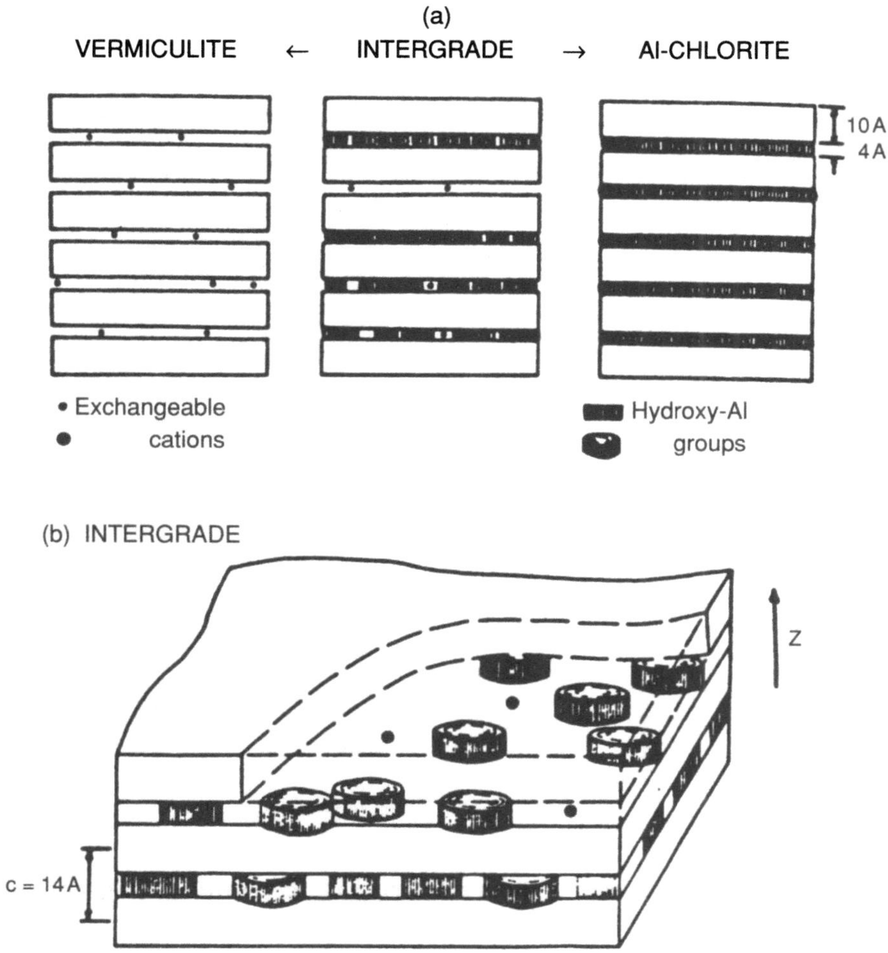

Fig. 3.2. Conceptual model of hydroxy-Al deposition in vermiculite in relation to vermiculite and chlorite end members. (Dixon and Jackson 1962)

In recent studies of the Mn oxides in soils, lithiophorite has frequently been identified (Uzochukwu and Dixon 1986; Golden et al. 1987). Two cases are in deep old red soils in well drained upland positions. The crystals have grown in the soil nodules to silt size, thus suggesting their formation in a favourable environment. This appears to be a third case of Al in the soil solution contributing to the formation of a relatively stable phase. But the growth of lithiophorite in neutral to calcareous soil horizons implies greater mobility of Al than formerly recognized in nonacid

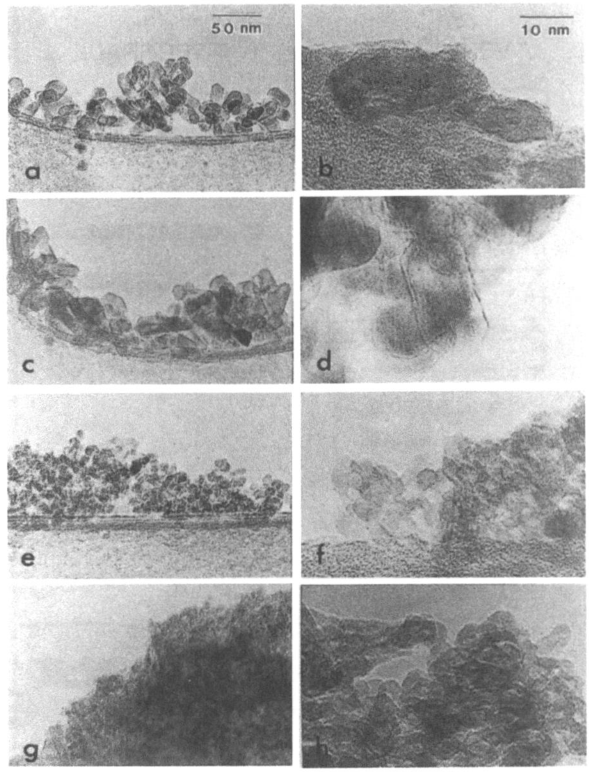

Fig. 3.3. The change in synthetic goethite crystal size with increasing mole% Al for Fe in the structure: **a, b** 0%; **c, d** 11%; **e, f** 24%; and **g, h** 34%. This figure is analogous to Dixon et al. 1983

soils. It also implies that hydroxy interlayers in vermiculite, generally recognized as a weathering product in acid surface soils, may form in nonacid soils too.

The formation of crystals in soils is an important indicator of the kinetics of soil development and of the chemical environment during their formation. The mixture of clay minerals in soil parent materials often precludes positive statements about mineral formation in such a complex environment. Yet, the growth of crystals in soil nodules provides localized conditions that are more reliably fixed in space and time than the typical soil clay. The growth of lepidocrocite sited earlier as an authigenic mineral in soil apparently has been joined by todorokite found recently in nodules in a Vertisol of Central America (Dixon et al. 1994). This todorokite exhibits trilling – a characteristic form of twinning of this mineral in natural and synthetic specimens (Fig. 3.4).

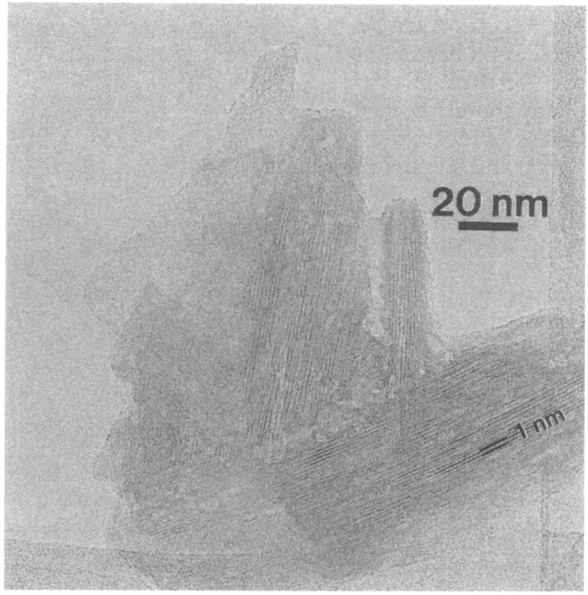

Fig. 3.4. Todorokite twins from nodules formed in a Vertisol of Guatemala. (Dixon et al. 1994)

3.9 Transient Soil Minerals

Some soils undergo frequent moisture changes, thus dissolution can occur on wetting and precipitation on drying. This is evident as soils dry in the summer and efflorescences of white gypsum and other salts form on the surface of the soil in, for example, a recently tilled garden soil that has dried in the sun for several hours. The particular salt that forms is a useful indicator of soil solution composition and may serve as a basis for evaluating important changes in the soil. For example, dispersion due to the presence of Na salts may be imminent and chemical and physical soil deterioration may follow. A yellow jarosite crystal efflorescence may indicate acidification and corrosive conditions for concrete and metal due to pyrite oxidation (Dixon et al. 1982).

Soils that are used for the production of rice in paddy culture are submerged for several months each year. They undergo drastic changes in water content seasonally. A recent study of a Vertisol of Texas revealed a red coating of iron oxide on the surfaces of shrinkage cracks and in pores (Fig. 3.5). Careful dispersion and fractionation of the red precipitates of this type permitted their identification as lepidocrocite and ferrihydrite by differential x-ray diffraction and electron microscopy (Wang et al.

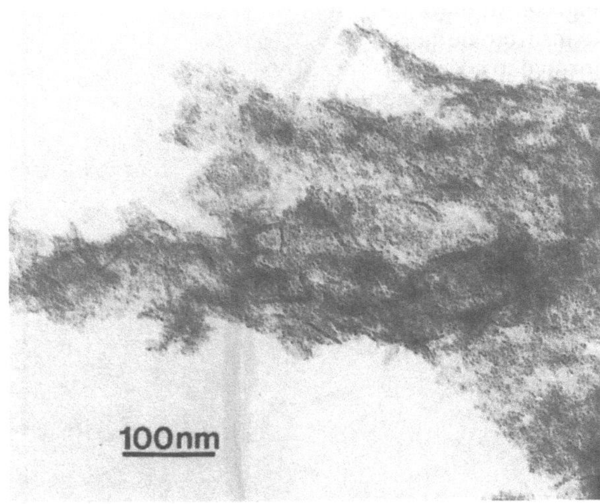

Fig. 3.5. Transient Fe oxide from Beaumont soil (Vertisol) in a rice paddy condition, identified as lepidocrocite (laths) and ferrihydrite (granules). (D.C. Golden and J.B. Dixon, unpub.; Wang et al. 1993)

Fig. 3.6. a Iron oxide-coated rice root and secondary root; **b** an Fe oxide cast of a cell in place with smooth exterior and rough interior surfaces. (Chen et al. 1980)

1993). These Fe oxide coatings bind soil particles together improving soil aggregation in a restorative process that occurs after the paddy is drained. These reactive Fe oxides were found to contain P and Si (Golden et al. 1988). The precipitation of orange-colored lepidocrocite iron oxide on the roots of rice plants illustrates a reaction widely viewed as a protective mechanism against Fe toxicity to the plant (Fig. 3.6).

3.10 Clay-Organic Interactions

The reactions between clays and organic substances range from cation exchange of simple organic cations to poorly understood interactions between clay surfaces and complex natural organic colloids. The pioneering research of Ensminger and Gieseking (1941) showed that organic cations are held on interlayer surfaces and influence the interlayer spacing of smectite, thus demonstrating the site of cation exchange. The advent of organic pesticides has stimulated many investigations of the reactions of these compounds with clays of soils and with model clay minerals. Clay minerals influence the fate of these substances. Thus, understanding clay organic reactions is of considerable environmental importance and is of importance in predicting the behavior of these substances and their effectiveness for pest control when applied to soils (Theng 1974).

Current restrictions on the movement of contaminants off a given site have stimulated the use of organic polymers in the western USA to aggregate soil in irrigation furrows to assure uniform water penetration where the use of full flooding may risk turbid return flow of irrigation water (Lentz et al. 1992). Special smectite-organic complexes (i.e. designer adsorbents) have been prepared for the enhanced adsorption of selected organic contaminants from waste water (Zielke et al. 1989).

3.11 Clays in Reclamation

When fresh rock is brought to the surface, any former equilibrium is disturbed and new weathering processes are begun. The reactions that begin first are likely to be the finer grained minerals and those that are the farthest from equilibrium in the new environment. The oxidation of Fe in chlorite is one of those reactions observed in the model oxidation of lignite overburden (Senkayi et al. 1981). The chlorite is converted to vermiculite and smectite via interstratified intermediates in laboratory experiments. Also, Fe oxides are precipitated. The exposure of siderite in mine spoil soon results in the formation of the Mn oxide, todorokite (Senkayi et al. 1986). The thin black todorokite covering masks the formation of Fe oxides that also occur during early siderite weathering. The exposure of Fe pyrite promotes oxidation with a resulting abrupt increase in acidity, and jarosite and gypsum formation may occur (Fig. 3.7a, b). Again, the Fe released is precipitated as Fe oxide (often with a jarosite intermediate stage). Thus, the red or brown color of Fe oxide is an important indicator of the depth of oxidation, the depletion of pyrite in lignite

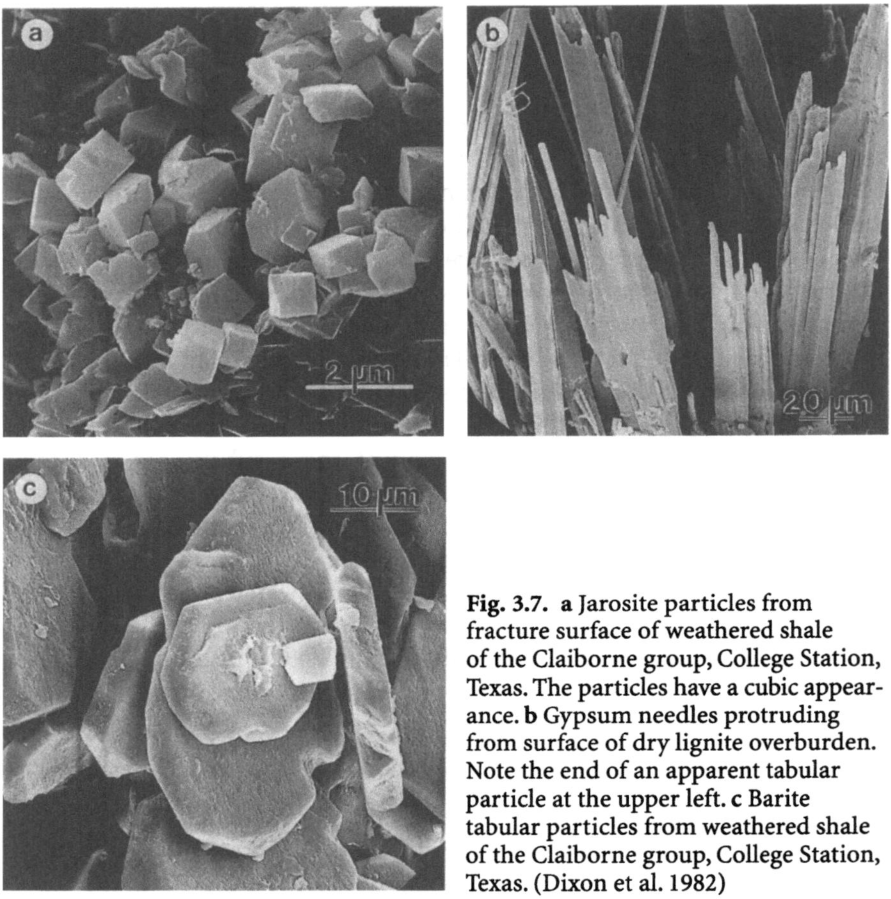

Fig. 3.7. **a** Jarosite particles from fracture surface of weathered shale of the Claiborne group, College Station, Texas. The particles have a cubic appearance. **b** Gypsum needles protruding from surface of dry lignite overburden. Note the end of an apparent tabular particle at the upper left. **c** Barite tabular particles from weathered shale of the Claiborne group, College Station, Texas. (Dixon et al. 1982)

overburden and the suitability of overburden for reclamation as soil material. The depth of oxidation also is a good indicator of the depletion of chlorite by weathering processes. Localized white barite occurs in soils associated with lignite overburdens in east Texas. Its origin is believed to be from Ba in feldspars in a volcanic ash component of the sediments (Fig. 3.7c).

3.12 Complementary Effects of Soil Minerals

While the diversity of minerals in soils adds considerably to the analytical problems involved in soil mineral investigations, it contributes to the quality of the soil. Soils that contain simple mineral suites tend to be poor and unproductive or abrupt and extreme in their physical and chemical properties. The Vertisols discussed earlier are examples of the latter. They contain too much smectite to give the desired versatility of a soil for most conventional uses. Acid smectitic soils are costly to improve because of their large component of reactive Al. The highly weathered soils of stable landscapes in warm humid climates have clays that are dominated by kaolinite and Fe oxides. Many of these soils are classified as Oxisols. They are low in native fertility and adsorb phosphate excessively (Fox 1980). On the other hand, oxisols, in spite of their high clay content, are receptive to water infiltration (El-Swaify 1980). The combination of kaolinite and Fe oxide particles of clay size produces a well-structured soil that is friable and porous. It should be recognized of course that good soils are not composed only of clay. The presence of coarser particles (silt and sand) adds considerably to their physical behavior and makes their management easier.

3.13 Clay Movement in Soils

Imogolite was first found in weathered glassy volcanic ash in Japan (Yoshinaga and Aomine 1962). More recent investigations have revealed that it occurs in spodosols developed from other parent materials in cool moist environments. Farmer et al. (1980) proposed that Al movement in podzolic soils is via proto-imogolite rather than via chelation as often suggested.

Fine clays are frequently recognized as mobile constituents in soils due to their depletion in surface layers and their accumulation below. The latter is particularly manifest in coatings of sand grains and soil aggregates (Fig. 3.8). Soils owe much of their friable behavior to the presence of porous aggregates that are bound by surfaces of relatively weak attachment, such as clay skins, to adjacent aggregates. These stand in contrast to the low porosity and compact particle arrangement of puddled or poorly structured soils.

3.14 Soil Taxonomy Reflects Clay Mineral Composition

Soil taxonomy provides a consensus on the roles of clays in soils. The USDA soil taxonomy system is based on field observations and laboratory data. Standard methods are documented and published, e.g. Soil

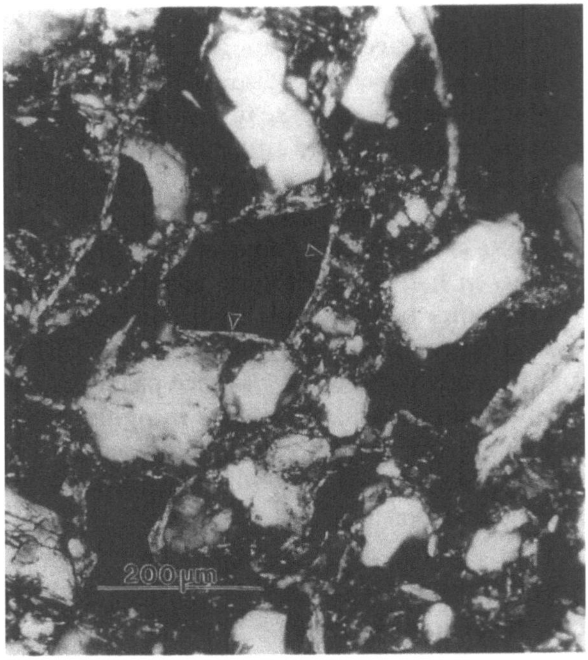

Fig. 3.8. Clay film on a skeletal grain (*arrows*) produced by the leaching and deposition of clay in the Gujranwala soil. (Alfisol of Pakistan; courtesy M. S. Akhtar)

Conservation Service (1984). Clay mineral composition is a part of that system. Soil taxonomy is a six-category system designed by a large group of collaborating scientists from many countries during a series of approximations. Soil Taxonomy (Soil Survey Staff 1975) had ten orders that were intended to include all soils of the world in the highest category of the system. As noted earlier, the 11th order has been added to accommodate Andisols due to the unique properties of allophane and imogolite they contain.

Two of the eleven soil orders call attention to soils frequently dominated by one or two clay minerals, Vertisols (smectites) and Andisols (allophane and imogolite).

3.15 Summary

Clays in soils offer a diversity of environments for investigation. Findings thus far show several departures from model minerals. With the advent of several new methods, soil mineralogy is likely to continue to provide challenging areas for study that will help explain and predict the many varied reactions of soils in their multiplicity of applications.

Acknowledgement. Christine Wallace is due special thanks for her excellent clerical and drafting services on this manuscript.

References

Allen BL, Hajeh BF (1989) Mineral occurrence in soil environments. In: Dixon JB, Weed SB (eds) Minerals in soil environments, 2nd edn. Soil Sci Soc Am, Madison, Wisconsin, pp 199–278

Arnalds O, Hallmark CT, Wilding LP (1995) Andisols from four different regions of Iceland. Soil Sci Soc Am J 59:161–169

Barnhisel RI, Bertsch PM (1989) Chlorites and hydroxy-interlayered vermiculite and smectite. In: Dixon JB, Weed SB (eds) Minerals in soil environments, 2nd edn. Soil Sci Soc Am, Madison, Wisconsin, pp 729–788

Bigham JM, Ciolkosz EJ (1993) Soil color. Soil Sci Soc Am Spe Publ No 31

Booltink HWG, Bouma J (1991) Physical and morphological characterization of bypass flow in a well-structured clay soil. Soil Sci Soc Am J 55:1249–1254

Chen CC, Dixon JB, Turner FT (1980) Iron coatings on rice roots: morphology and models of development. Soil Sci Soc Am J 44:1113–1119

Chen CC, Turner FT, Dixon JB (1989) Ammonium fixation by high-charge smectite in selected Texas Gulf Coast soils. Soil Sci Soc Am J 53:1035–1040

Dixon JB (1994) Persistent and transient minerals in Vertisols. In: Transactions of 15th world congress of soils science, Acapulco, Mexico, vol 9 (Suppl), pp 467–478

Dixon JB, Jackson ML (1962) Properties of intergradient chlorite-expansible layer silicates of soil. Soil Sci Soc Am Proc 26:358–362

Dixon JB, Hossner LR, Senkayi AL, Egashira K (1982) Mineralogical properties of lignite overburden as they relate to minespoil reclamation. Acid Sulfate Weathering. Soil Sci Soc Am Spec Publ 10:169–191

Dixon JB, Golden DC, Calhoun FG, Buseck PR (1983) Synthetic aluminous goethite investigated by high-resolution transmission electron microscopy. In: Bailey GW (ed) Proc 41st annual meeting of electron microscopy society of America, Phoenix, Arizona, pp 192–193

Dixon JB, Jacob JS, White GN (1994) Todorokite in manganese oxide nodules of Guatemalan Vertisol. Soil Sci Soc Am Annual Meeting, Seattle, pp 409–410

Dudal R, Eswaran H (1988) Distribution properties and classification of Vertisols. In: Wilding LP, Puentes R (eds) Vertisols: their distribution properties, classification and management, Texas A&M University Printing Center, College Station, 1–22

El-Swaify SA (1980) Physical and mechanical properties of Oxisols. In: Theng BKG (ed) Soils with variable charge. New Zealand Society of Soil Science, c/- Soils Bureau, Department of Scientific and Industrial Research, Private Bag, Lower Hutt, pp 303–324

Ensminger LE, Gieseking JE (1941) The adsorption of protein by montmorillonite clays and its effect on base exchange capacity. Soil Sci 51:125–132

Farmer VC, Russell JD, Berrow ML (1980) Imogolite and proto-imogolite allophane in spodic horizons: evidence for a mobile aluminum silicate podzol formation. J Soil Sci 31:673–684

Fox RL (1980) Soils with variable charge: agronomic and fertility aspects. In: Theng BKG (ed) Soils with variable charge. New Zealand Society of Soil Science, c/- Soils Bureau, Department of Scientific and Industrial Research, Private Bag, Lower Hutt, pp 195–224
Golden DC, Chen CC, Dixon JB (1986) Ion exchange, thermal transformations, and oxidizing properties of birnessite. Clays Clay Miner 34:511–520
Golden DC, Dixon JB, Kanehiro Y (1987) Properties of lithiophorite in an Oxisol from Hawaii. Agronomy Abstracts, Atlanta, Georgia, 273 pp
Golden DC, Dixon JB, Sittertz-Bhatkar H, Turner FT (1988) Seasonally precipitated iron phases in Beaumont paddy soil of east Texas. Agronomy Abstracts, Anaheim, California, p 307
Jones RC (1981) X-ray diffraction line profile analysis vs. phosphorus sorption by 11 Puerto Rican soils. Soil Sci Soc Am J 45:818–825
Jones DE, Jones KA (1987) Treating expansive soils. Civil Engineering. Am Soc Civil Eng 57:62–65
Leamy ML, Smith GD, Colmet-Daage F, Otowa M (1980) The morphological characteristics of Andisols. In: Theng BKG (ed) Soils with variable charge. New Zealand Society of Soil Science, c/- Soils Bureau, Department of Scientific and Industrial Research, Private Bag, Lower Hutt, pp 17–34
Lentz RD, Shainberg I, Sojka RE, Carter DL (1992) Preventing irrigation furrow erosion with small applications of polymers. Soil Sci Soc Am J 56:1926–1932
MacEwan DMC (1950) Some notes on the recording and interpretation of X-ray diagrams of soils clays. J Soil Sci 1:90–103
McKenzie RM (1989) Manganese oxides and hydroxides. In: Dixon JB, Weed SB (eds) Minerals in soil environments, 2nd edn. Soil Sci Soc Am, Madison, Wisconsin, pp 439–465
Ming DW, Mumpton FA (1989) Zeolites in soils. In: Dixon JB, Weed SB (eds) Minerals in soil environments, 2nd edn. Soil Sci Soc Am, Madison, Wisconsin, pp 873–911
Norrish K, Taylor RM (1961) The isomorphous replacement of iron by aluminium in soil goethites. J Soil Sci 12:294–306
Pearson RW, Ensminger LE (1949) Types of clay minerals in Alabama soils. Soil Sci Soc Am Proc 13:153–156
Randolph JW, Reed IF (1938) Test of tillage tools II – effects of several factors on the reactions of fourteen-inch moldboard plows. Agro Eng 19:29–33
Ritchie JR, Kissel DE, Burnett E (1972) Water movement in undisturbed swelling clay soil. Soil Sci Soc Am Proc 36:874–879
Robert M (1987) Potassium availability in relation to soil minerals. Trans XIII Congr Int Soil Science, Hamburg 13–20 Aug 1986. c/o Dr. Hintze, Ordinariat für Bodenkunde, Allende Platz, 2000 Hamburg 13, Germany, 1121 pp
Sawhney BL (1972) Selective sorption and fixation of cations by clay minerals: a review. Clays Clay Miner 20:93–100
Schwertmann U, Taylor RM (1989) Iron oxides. In: Dixon JB, Weed SB (eds) Minerals in soil environments, 2nd edn. Soil Sci Soc Am, Madison, Wisconsin, pp 379–438
Senkayi AL, Dixon JB, Hossner LR (1981) Transformation of chlorite to smectite through regularly interstratified intermediates. Soil Sci Soc Am J 45:650–656
Senkayi AL, Dixon JB, Hossner LR (1986) Todorokite, goethite, and hematite: alteration products of siderite in east Texas lignite overburden. Soil Sci 142: 36–41

Soil Conservation Service (1984) Procedures for collecting soil samples and methods of analysis for soil survey. Soil Survey Investigations Report 1, United States Department of Agriculture, Lincoln, Nebraska, 68 pp
Soil Survey Staff (1975) Soil taxonomy: a basic system of soil classification for making and interpreting soil surveys. Soil Conservation Service, United States Department of Agriculture, Agriculture Handbook No 436. US Government Printing Office, Washington DC 20402, 754 pp
Soil Survey Staff (1990) Keys to soil taxonomy, 4th edn. Soil Manage Support Serv Tech Monogr 19, Virginia Polytechnic Institute and State University, Blacksburg, Virginia
Southard RJ, Graham RC (1992) Cesium-137 distribution in a California pelloxerert: evidence of pedoturbation. Soil Sci Soc Am J 56:202–207
Theng BKG (1974) The chemistry of clay organic reactions. John Wiley, New York, 343 pp
Uzochukwu GA, Dixon JR (1986) Mineralogical and chemical properties of black concretions in two soils of east Texas and north Alabama. Soil Sci Soc Am J 50:1358–1363
Wada K (1989) Allophane and imogolite. In: Dixon JB, Weed SB (eds) Minerals in Soil Environments, 2nd edn. Soil Sci Soc Am, Madison, Wisconsin, pp 1051–1087
Wang HD, White GN, Turner FT, Dixon JB (1993) Ferrihydrite, lepidocrocite, and goethite in coatings from east Texas vertic soils. Soil Sci Soc Am J 57:1381–1386
Warkentin BP, Maeda T (1980) Physical and mechanical characteristics of Andisols. In: Theng BKG (ed) Soils with variable charge. New Zealand Society of Soil Science, c/- Soils Bureau, Department of Scientific and Industrial Research, Private Bag, Lower Hutt, pp 281–301
Yoshinaga N, Aomine S (1962) Imogolite in some Ando soils. Soil Sci Plant Nutr 8(3):22–29
Zielke RC, Pinnavaia TJ, Mortland MM (1989) Adsorption and reactions of selected organic molecules on clay mineral surfaces. In: Sawhney BL, Brown K (eds) Reactions and movement of organic chemicals in soils. Soil Sci Soc Am Spec Publ 22:81–97

Chapter 4

Interactions of Non-Volatile Micro-organic Pollutants and Clay Minerals in Surficial Environments

W. A. House

4.1 Introduction

Micro-organic compounds are ubiquitous in aquatic environments, occurring both in solution and in association with particles and colloids. Many of the compounds are of anthropogenic origin and are classed as pollutants although not toxic at the concentrations that are commonly encountered. Their transport and persistence in the environment are complex topics needing detailed information about the compound's specific chemical structure, biochemistry and properties, as well as the solution composition. In general, the fate of the compounds is determined by a combination of parameters including vapour pressure, volatility, aqueous solubility, chemical/biochemical degradation and interaction with solid surfaces. It has been known for some time that organic compounds can interact with clay minerals, and be involved in both ion-exchange reactions and also penetration of the interlayer space of expandable clays (Rausell-Colom and Serretosa 1978).

The main emphasis, concerning the interaction or partition of micro-organics with surfaces, has been to find simple regression equations that enable a prediction of the distribution of the compound between the solution and solid phase, based on some easily measured property of the compound and the sediment. Traditionally this distribution has been quantified through a distribution or partition coefficient, K_d (see Sect. 4.2.3.1). This expresses the relative concentration of the compound on the solid phase with respect to the solution concentration, or more correctly the concentration in the filtrate, i.e. that passing through a 0.45-μm membrane filter. The most common predictors are the octanol-water partition coefficient of the compound and the organic carbon or organic matter content of the sediment. Hence, from a knowledge of these two parameters for the system in question, it is possible to estimate the distribution of the compound between the aqueous and solid phases when at equilibrium. This approach is, needless to say, only a first approximation, its development being driven by the need for simple tools which can be applied to different systems including complex sediments or soils. If the interactions with sediment are not dominated by hydrophobic mechanisms, or the structure of the organic material is very variable, then the regression relationships are not likely to be appropriate. For example, many herbicides adsorb strongly to both clay minerals and organic carbon components in soil, and any prediction of herbicide movement with particulates must in general examine the relative importance of these associations (Grundl and Small 1993).

In general, individual micro-organic compounds have specific types of molecular interactions with sediment components. These components may be very varied, with different mineral and organic material influencing the overall interaction of the compounds and also the kinetics of the interaction. Hence, even in situations with hydrophobic neutral compounds, the chemical nature of the organic material associated with the particle surface may influence the adsorption affinity. Even in soils where this appears to occur, the removal of the organic material by sodium hypochlorite still leaves surfaces which are capable of substantial adsorption, e.g. sorption of trifluralin and atrazine (Francioso et al. 1992). In some situations, e.g. clay soils with a low organic matter content, the interactions of the micro-organic compounds will be determined by the interactions with the high specific surface area clays; the subsequent movement of the organic compounds into drainage channels and rivers will therefore be influenced by these interactions.

Unfortunately, there is little information about the role of clays in the transport of micro-organic pollutants, particularly in river sediments. Most of the research has been limited to looking for correlations between the concentration of specific compounds in sediments and soils and some property of the particulates such as the cation exchange capacity (cec), organic carbon (OC) organic matter (OM) content, or clay content based on size fractionation ($<2\ \mu m$). However, from laboratory experiments, particularly with expandable clays, there is ample evidence to indicate the importance of clay surfaces and interlayers both in terms of their affinity and capacity for some compounds and also their ability to influence the release and uptake kinetics.

In rivers and lakes, the main sources of minerals, including clays, are: (1) internal cycling from bed sediments which is mainly induced by mass and thermal convection, wind action or bioturbation; (2) subsurface or surface runoff from the catchment which is influenced by land management practices in the region; (3) aerial wet or dry deposition; and (4) bank erosion, particularly during spates. In these situations the clay may have been in contact with micro-organic compounds, depending on the location and the history of the site. For material entering a river or lake, the particulates will be exposed to natural organic matter (NOM) composed of many organic compounds which will compete for surface adsorption sites, and the surface will develop in different ways depending on the locality.

It is impossible in this chapter to cover the many thousands of micro-organic compounds and multitude of clays (see for example Velde 1992). As noted above, each compound has particular properties affecting the mechanisms of their interaction with clays and broad generalisations are rarely possible. Most of the research reported in the literature has focused on particular groups of compounds, e.g. anionic surfactants, triazine and triazole herbicides and the quats, and most studies of clays have used kaolinite or montmorillonite as the adsorbent. Hence, this chapter will reflect this emphasis although other systems will be mentioned when appropriate. The chapter covers important aspects of the complex composition of sediments and the key factors which influence molecular interactions of micro-organic compounds with clay minerals. Following this, the kinetics of the interaction and adsorption behaviour are discussed in terms of model isotherms and the possibilities of interpretation of the isotherm data by various sorption mechanisms. These mechanisms are discussed for a range of adsorbent/adsorbate combinations, with examples of mechanisms discussed for basic and acidic compounds. Finally, a case study is presented concerning the interaction of a group of pesticides with sediments from a river catchment with particular emphasis on the role of the clay components in the sediment on the adsorption behaviour of the sediment.

4.1.1 Sediment Composition

Natural sediments and soils are usually a complex mixture of minerals, natural organic matter (NOM) and biological components. The minerals may consist of a variety of components including clays, carbonates and quartz. In addition, the composition of the surface of the mineral components is influenced by its exposure to the aquatic environment, either in soil systems or, for example, in rivers and lakes. This invariably leads to complex interfaces with coatings of solid precipitates, coprecipitates or biologically derived material. For example, the coprecipitation of inorganic phosphate with calcite occurs in hardwater lakes (Kuchler-Krischun and Kleiner 1990), with the possibility of the formation of distinct calcium phosphate minerals at high phosphate concentrations. In soft waters, coatings of Fe and Mn-oxides have been reported and are thought to

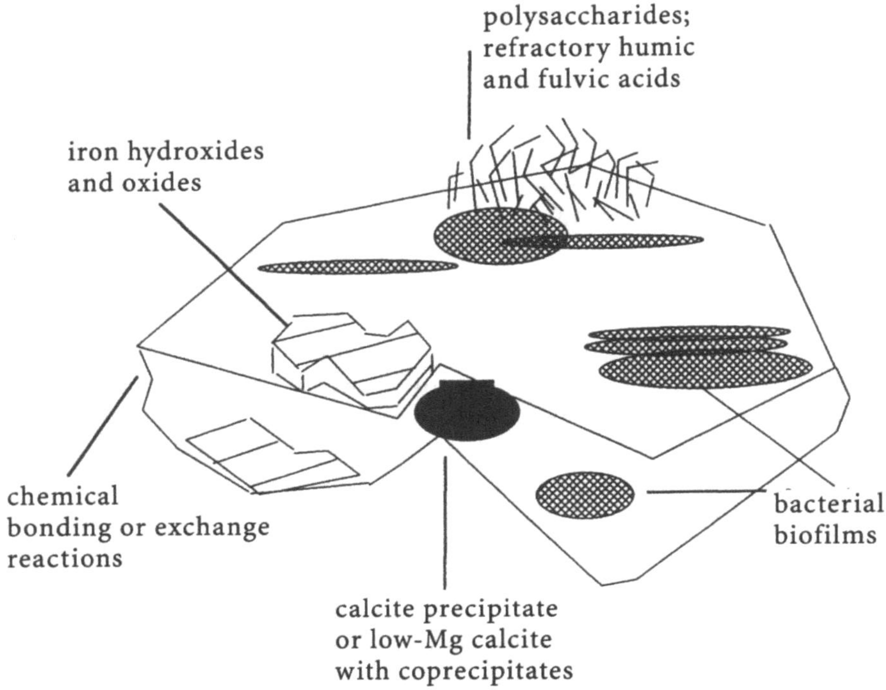

Fig. 4.1. A schematic of a complex surface of a particle aggregate after exposure to a natural water. The surface may develop heterogeneous regions through, for example, precipitation reactions of carbonates or metal oxides/hydroxides, biofilm formation initiated by the attachment of bacteria and exchange or adsorption reactions of solutes or polymeric materials such as polysaccharides and refractory fulvic acids

influence the surface reactions, e.g. the reaction of inorganic phosphate with iron hydroxide to form amorphous solid solutions of iron hydroxide and phosphate (Fox 1991). In water originating from catchments containing peat, the clay particles may interact with humic and fulvic acids (Tipping 1990) and therefore influence the surface charge and interaction with dissolved metals such as Fe(II) and Mn(II). Therefore, coatings on surfaces can form through a multitude of processes including: (1) mineralisation reactions noted above, (2) sorption of low molecular weight molecules, (3) flocculation or sorption of organic macromolecules such as acidic polysaccarides or humic acids and (4) bacterial exudates formed through bacterial colonisation of aggregated or large particles as well as cellular material from bacteria and benthic algae (Costerton et al. 1994). An example of this complexity is depicted in Fig. 4.1 for a hypothetical clay aggregate.

4.1.2 Clay-Solution Interface

The surface of a pure clay mineral in a suspension of a simple electrolyte is largely determined by the ionic reactions of the surface groups and the structure of the double-layer at the solution-solid interface. The interface may be regarded as consisting of a surface plane or 'O' plane where inner-sphere complexes (or surface complexes) are formed, e.g. the reaction between alkaline earth metals and surface silanol groups, a region further from the surface up to the inner Helmholtz plane or β-plane, where outer-sphere complexes are centred, and finally the outer Helmholtz plane marking the start of the diffuse layer. The chemical and ionic interactions in the 'O' plane involve dehydrated or partially dehydrated ions and influence the surface potential and charge density at the surface. At a distance further from the surface, the counter-ions which bind with the oppositely charged surface groups and form the outer-sphere complexes are thought to retain their water of hydration. These ions are mainly bound through electrostatic interactions and have a lower bonding energy than the inner-sphere complexes. Still further from the surface, in the region of the hydrodynamic shear plane, the ions are only influenced by an average electrostatic field and their density follows a Boltzmann distribution, influenced by their normal thermal motion and electrostatic forces. This region is referred to as the diffuse layer and has a corresponding surface potential and charge density associated with it (Everett 1992).

An example of the interfacial region for kaolinite in a NaCl solution is depicted in Fig. 4.2. The double layer is only shown for the "broken edge sites", where the surface charge is variable depending on the pH of the solution and surface complexation reactions. In summary, three main surface structures can be identified: a gibbsite surface with a pK ($pK = -\log K$) of approximately 4, similar to gibbsite (Table 4.1), a siloxane surface and the crystal edge with a combination of SiOH and AlOH sites with $pK = 6.3$ (Wieland et al. 1988). The negative charge of the planar surface is mainly determined by the isomorphous substitution of Si by Al in the siloxane layer. Usually the edge sites contribute $< 20\%$ of the surface area of the crystals and so, irrespective of the pH, the negative charge on the face of the platelets is greater than the positive edge charge. The basal hydroxyls are very poor proton donors and may form hydrogen bonds with only very strong bases such as hydrazine. The siloxane surface is a very poor electron donor and forms weak hydrogen bonds with strong proton donors. The oxygen plane has a constant negative charge because of the occasional isomorphous replacement of Si by Al in tetrahedral sheets. The exchangeable cations, both inorganic ions as well

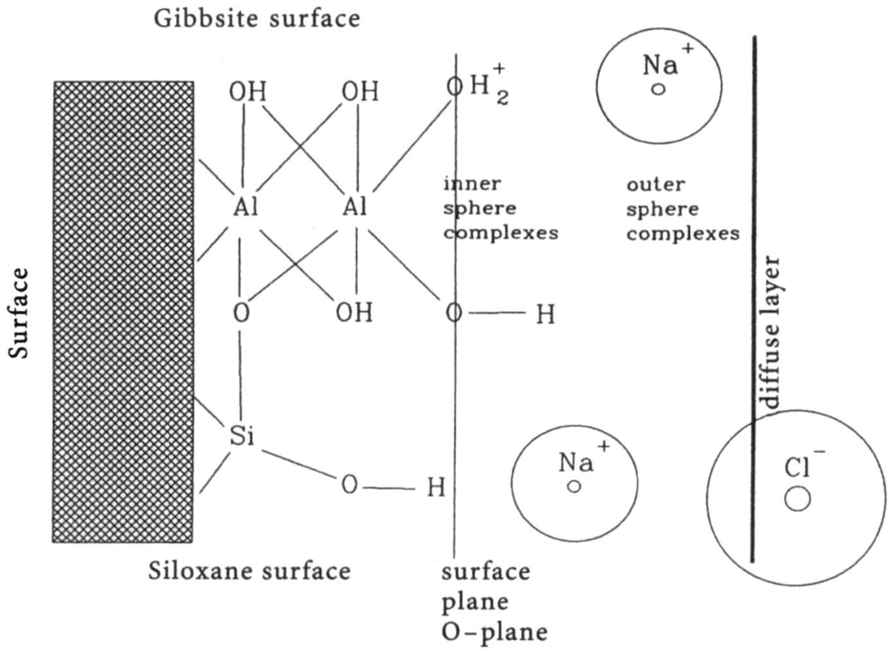

Fig. 4.2. Schematic of the electrical interface at the edge surface of kaolinite at neutral pH. The gibbsite and siloxane surfaces are indicated

as charged organic compounds (e.g. quaternary ammonium surfactants (QACs) and herbicides such as paraquat and diquat), may compensate for the charge distributed over the plane of the silicate.

Smectite clays adsorb either at permanent-charge sites on the interlayer basal planes or at edge sites. Although the interlayer sites are more abundant than the edge sites, the edge sites may determine the adsorption characteristics of the clay depending on the solution conditions. This has been demonstrated for the adsorption of Co(II) (0.1 mM) on montmorillonite. At low pH (<7) and sodium ion concentration of 0.1 mmol dm^{-3}, sorption is mainly on the interlayer sites. The edge sites become of increasing importance as the pH increases to 10. With high sodium concentrations (> 0.1 mol dm^{-3}) and pH > 7, the clay edges are the dominant adsorption sites as a result of the competition for interlayer sites by Na^+ and increased ionisation of the silanol groups on the edge sites with increasing pH (Papelis and Hayes 1996).

Information about the electrical properties of the interface of clays is of vital importance in the consideration of their interaction with organic solutes (Table 4.1). In the case of kaolinite, for example, anionic com-

Table 4.1. Surface ionisation reactions for various surfaces relevant to the study of clays; the pKs (pK = –log K) are the acid dissociation constants chosen from the literature

Reaction	Mineral surface	pK	Reference
$AlOH_2^+ = AlOH + H^+$	Kaolinite Basal gibbsite	4.04	Schindler and Stumm (1987)
$AlOH = AlO^- + H^+$	ibid	7.04	Wehrli et al. (1990)
$MOH_2^+ = MOH + H^+$	Kaolinite SiOH/AlOH edge	6.31	Schindler and Stumm (1987)
$MOH = MO^- + H^+$	ibid	8.69	ibid
$SiOH_2^+ = SiOH + H^+$	Silica/quartz	2.30	ibid
$SiOH = SiO^- + H^+$	Silica/quartz	7.30 7.54	House and Orr (1992)
$AlOH_2^+ = AlOH + H^+$	Gibbsite Edge sites	10.00	Hiemstra et al. (1987)
$AlOH_2^+ = AlOH + H^+$	γ-Al_2O_3	7.40	Kummert and Stumm (1980)
$AlOH = AlO^- + H^+$	γ-Al_2O_3	10.00	ibid

pounds such as sodium dodecyl sulphate interact in acid and neutral solutions with the positively charged platelet edges, whereas cationic compounds, such as cetylpyridium chloride, absorb to the cleavage surface which has a permanent negative charge. As discussed later (Sect. 4.3.2), the interactions may also be influenced by hydrophobic interactions which are less easily characterised.

4.1.3 Micro-organic Properties

Having discussed the properties of the clay interface it is relevant to briefly mention some of the molecular properties of the adsorbates which influence their interaction with clay surfaces. These properties include:

1. The solubility of the compound in solution and its dependence on temperature, pH and ionic strength. This property largely determines the mass transport of the compound and its distribution in the aquatic environment.

2. The octanol-water partition coefficient (K_{ow}), which is a simple measure of the hydrophobic character of the molecule and has been used extensively to predict the partitioning of pesticides with natural organic matter (NOM) associated with sediments and soil. K_{ow} is defined as the ratio of the concentration of the compound in octanol and in water in a mixed binary solution.
3. The basicity of the compound, defined through the dissociation constant, K_B:

$$B + H_2O = BH^+ + OH^-$$
$$K_B = a_{BH^+}\, a_{OH^-}/a_B \qquad (1)$$
$$pK_B = pOH + \log\,[a_B/a_{BH^+}],$$

where B represents the base. When $pK_B = pOH$, $a_B = a_{BH^+}$. Weakly basic compounds have high pK_Bs, indicating they are poor proton acceptors, e.g. urea with a $pK_B = 13.82$ has a lower basicity than aminotriazole with a $pK_B = 9.83$.
4. The acidity of the compound is correspondingly defined through the dissociation of the acid:

$$BH^+ + H_2O = H_3O^+ + B$$
$$K_A = a_{H^+}\, a_B/a_{BH^+} \qquad (2)$$
$$pK_A = pH + \log\,[a_{BH^+}/a_B],$$

where K_A is the acid dissociation constant related to pK_B by: $pK_B = pK_w - pK_A$, where K_w is the ionic product of water. The higher the pK_A value of the compound, the weaker the acidic property. The degree of ionization of the molecule is therefore determined by the thermodynamic dissociation constant and the ionic strength of the solution. As discussed below, the clay surface may produce large perturbations of the solution conditions from those in the bulk of solution and that measured using conventional techniques.

4.2 Isotherms

4.2.1 Measurement of Sorption/Desorption Isotherms

The methods generally used to measure solute adsorption to minerals are also suitable for isotherm determinations in micro-organic solute/clay systems. These include:

1. The so-called batch method, in which a known mass of clay is put in contact with a solution of the micro-organic compound and left for a pre-determined period before the aqueous concentration of the compound is measured in the supernatant. The period of contact is either found from kinetic measurements of the sorption process, or if the kinetics are very slow it is usually some arbitrary time, e.g. 24 h. The separation of the solution from the suspension at the end of the sorption period is a critical step in the method as the suspension temperature must remain constant during centrifugation and/or filtration. The handling of micro-organic compounds usually presents particular problems caused by the interaction of the compounds with containing vessels and filters, and contamination from plastics or other material with which the solutions come into contact (House and Ou 1992). Such contaminants can interfere with the sorption measurement either directly, through their competition for the adsorbent surface with the compound of interest, or indirectly by their interference in the analytical method.
2. The use of an automated cell to measure the isotherm on a single batch of clay, e.g. House and Farr (1989). In method (1) each isotherm point corresponds to a separate batch of adsorbent whereas in method (2), the adsorption and desorption isotherms are determined on a single batch of adsorbent. The adsorbate is added by sequential addition after a predetermined contact time. The method is convenient to automate and permits easy control of the solution conditions such as the dissolved oxygen concentration, pH and temperature. The cell is also suited for measurements of the kinetics (>20 min) of sorption and desorption, particularly slow processes taking >12 h. Again, the handling of the micro-organic compounds needs special precautions to avoid sorption to container materials and the filter membrane. With some compounds, such as the synthetic pyrethroids, their interaction with the filter membrane is unavoidable but can be minimised by an appropriate choice of membrane material and measured by operating the cell without the adsorbent, i.e. measuring the sorption properties of the cell including the membrane.

In both methods, the design of the experiment is of crucial importance for reliable sorption/desorption measurements. The choices of the mass of adsorption and mass of the micro-organic compound in the adsorption vessel are particularly important. For example, if the compound is only weakly adsorbed to the clay, enough clay must be present in suspension to produce a large enough change in the concentration of the adsorbate in solution to meet the necessary analytical requirements. Failure to do this may lead to the erroneous conclusion that the micro-organic compound is not adsorbed. In contrast, when the adsorption affinity of the substrate is high, a situation could arise where the micro-organic compound is removed from solution, producing a concentration below the limits of determination of the compound. In this case either the initial concentration (or volume of solution) of the compound must be increased or the mass of the clay decreased. Inevitably, with an unknown sorption system, it is necessary to make preliminary measurement of the sorption affinity using different suspended solids concentrations at a fixed initial adsorbate concentration.

Normally, for measurements with micro-organic compounds in a concentration range of relevance to natural waters, the sorption isotherm is linear. In this situation, an estimate of the mass of adsorbent and volume of suspension may be made to achieve 50% sorption of the compound given an estimate of the distribution coefficient or adsorption constant for the system of interest. Constructing a mass balance of the initial amount of the micro-organic compound in solution prior to sorption and the sum of the amount sorbed and in solution after equilibrium leads to:

$$V\,c_i = K_d\,c_e\,w + c_e\,V, \tag{3}$$

where V is the solution volume (ml), c_i the initial concentration of the micro-organic in solution (µg ml^{-1}), K_d is the distribution coefficient – see Eq. (7) below – in ml g^{-1}, c_e is the equilibrium concentration of the compound (or concentration at the end of the contact time) in mg ml^{-1} and w is the mass of the clay adsorbent in grams. It follows from Eq. (1) that the equilibrium concentration is:

$$c_e = V\,c_i/(K_d\,w + V), \tag{4}$$

and the percentage sorption of the compound is:

$$\text{percentage sorption} = K_d\,w/(K_d\,w + V). \tag{5}$$

An added complication for micro-organic sorption measurements is the possibility of degradation of the compound during the measurements. The degradation may be either chemical (e.g. alkaline hydrolysis, photochemical) or biologically mediated. Although the effect of these processes, relative to sorption, may not be important at high solute concentrations, the same is not true for the trace concentrations often studied. In these situations, caution is necessary to avoid including the degraded component as part of the calculated "sorption amount". Similarly, if the compound is volatile, special procedures are necessary to measure the sorption isotherm. If degradation of the micro-organic compound is suspected, the concentration of the degradation product(s) must be measured at the end of the contact time. It is also normally good practice to separate the solid adsorbent after sorption, and directly measure the concentration of the micro-organic compound sorbed to enable a comparison with the amount calculated from the change in the solution concentration. This is often a lengthy procedure involving soxhlet or critical-fluid extraction followed by a solvent concentration stage and appropriate chromatography.

4.2.2 Kinetics of the Interaction

This is a complex topic in its own right and the reader is referred to reviews (e.g. Calvet 1989). As noted above, an initial study of the kinetics of sorption or desorption is essential before isotherms are measured. A study of the kinetics and temperature dependence, provide extra information about the mechanism of the surface interaction.

Many compounds exhibit two phases of kinetics, with an initial fast adsorption stage lasting from a few minutes to several hours followed by a slow stage of many hours to several days. Similar behaviour is also shown for some micro-inorganics such as inorganic phosphate. The initial phase is often reversible and termed "labile adsorption" (Gilchrist et al. 1993) and is easily extracted with water or organic solvents, i.e. in this instance the fraction is operationally defined. The slower stage is associated with the molecules diffusing in the solid matrix, either into the NOM associated with natural particles, intra-particle diffusion (Wu and Gschwend 1986) or interlameller spacing of expandable clays. The fraction associated with the slow kinetics may be termed the "bound" fraction because it is less easily released to water and organic solvents, e.g. the quats in the interlayer spacing of clays. There is now evidence that the triazines are sorbed to soils by processes in which fast and slow kinetics play a part, so that part of the adsorbed triazine is released quickly into water whereas a more recalcitrant fraction is released more slowly. The length of time for which compounds are exposed to soils in field conditions also influence the relative amounts of the labile and bound fractions. The precise mechanism of adsorption in complex sediments and soils is still open to debate. Diffusion models may be used to describe the slow kinetic stage (Wu and Gschwend 1986). The results indicate that larger particles have slower uptake kinetics with compounds with a higher K_{ow} showing the slowest kinetics. This is consistent with diffusion of molecules through the pore fluid held in the interstices of natural aggregates. The movement of the adsorbate is retarded by its interaction with the surfaces in the internal surface in the aggregates.

The fast kinetic stage may be described by a first-order rate equation of the form:

$$d\,n_a/dt = k_1\,[K_d\,C - n_a]\,, \tag{6}$$

where C is the concentration of the micro-organic compound in solution, n_a is the sorption amount, t is time, k_1 the first-order rate constant and K_d the distribution coefficient. Other rate equations have also been applied including the Freundlich two-site equation, parabolic diffusion law (Kookana et al. 1992) and various black box models (Wu and Gschwend 1986). However, there are few detailed studies of the kinetics of sorption to clay minerals in the absence of NOM.

Most studies of the sorption kinetics have utilised the batch method discussed above. In fact, the adsorption and desorption kinetics generally depend on the solution hydrodynamics at the aggregate/water interface. For instance, adsorption of micro-organic compounds to a river bed sediment will be different from the same sediment in suspension; similarly for water flowing over soils (Kookana et al. 1992). However, no information is available about the effects of flow on the sorption kinetics and the role of clays in this interaction is unknown.

4.2.3 Description of the Sorption Isotherm

It is important to summarise sorption/desorption isotherms to provide information about the temperature dependence, sorption affinity and sorption capacity of the system under consideration, and to allow comparisons between different micro-organic/clay systems.

4.2.3.1 *Linear Adsorption Isotherm*

Most isotherms for micro-organic/sediment systems are linear, viz:

$$n_a = K_d\,c_e\,, \tag{7}$$

where K_d is the distribution coefficient (sometimes referred to as the partition coefficient), n_a is the sorption amount expressed in the amount sorbed per unit mass of dry clay (e.g. $\mu mol\ g^{-1}$) and c_e is the equilibrium concentration of the compound. If an equilibrium is not achieved because of the nature of the sorption/desorption kinetics, c_e is the concentration after the time of contact, e.g. c_{24} would be the concentration of the compound after 24 h of contact with the clay.

Generally, all isotherms become linear in the limit of low adsorbate concentrations (see below for the Langmuir and Freundlich equations). If the surface is heterogeneous, possessing regions with varying affinities for the adsorbate according to Eq. (7), the isotherm is still linear, with the measured K_d effectively the sum of the K_ds for the component regions. A linear isotherm is expected for surfaces possessing surplus adsorption sites so that the surface adsorbate density is low, and there is no interaction between the adsorbed molecules leading to lateral surface interactions, surface saturation or co-operative adsorption effects (see Sec. 4.3.2).

4.2.3.2
Langmuir Isotherm

This is a semi-empirical equation widely used to describe adsorption in both gas-solid and liquid-solid systems. In contrast to the linear isotherm, the Langmuir equation does account for the finite number of sorption sites on the surface, but its theoretical derivation necessitates the assumption that the sorption sites are identical and that the lateral interactions on the surface can be neglected.

$$n_a = n_m K_l c_e/(1 + K_l c_e), \tag{8}$$

where K_l is the Langmuir constant, n_a is the sorption amount, c_e the equilibrium concentration and n_m the monolayer capacity or amount sorbed at surface saturation. From Eq. (8), a linear isotherm is found for $K_l c_e \ll 1$. The adsorption constant, K_l, is a measure of the sorption affinity (as K_d above), the temperature dependence of which enables the heat of adsorption to be calculated.

4.2.3.3
Freundlich Isotherm

This is also a semi-empirical equation widely used in soil science. The isotherm equation takes the form:

$$n_a = K_f c_e^{1/n}, \tag{9}$$

where K_f and n are the Freundlich adsorption constants which can be calculated from a regression analysis of the logarithmic form of Eq. (9). With n = 1, a linear isotherm is obtained.

4.3 Sorption Mechanisms

The adsorption of micro-organic compounds generally depends on several factors including the chemical nature and preparation of the clay, the ionic and general composition and pH of the solution, and the chemical character of the adsorbate. Early work investigating the interaction of a range of basic and acidic compounds with montmorillonite demonstrated the complexity of the interactions but noted several general conclusions including for example (Bailey et al. 1968):

1. Sorption on highly acidic H-montmorillonite was greater than on the near-neutral Na-montmorillonite.
2. It is clear that the acidic or basic character of the adsorbate is a fundamental property influencing the adsorption affinity. The interaction of basic compounds depends on the density of edge and interlayer acid sites (for expandable clays) and their acid strength as well as the basicity of the adsorbate.
3. Aqueous systems produced isotherms in general agreement with the Freundlich equation rather than the Langmuir equation.

Conclusion (2) leads to some fundamental problems in predicting the adsorption behaviour in some systems. Although there is usually information about the acid-base properties of many micro-organic compounds in bulk solution, information is not readily available about the acid/base properties of the edge sites and interlayer spacing of the clays. for example, several investigators have noted the difference between the bulk solution pH and the pH at the surface or internal surface of the clay (Bernstein 1960). This difference arises from the ionisation of the clay surface and the development of a diffuse double layer. The hydrogen ion activity at the surface ('O' plane in Fig. 4.2), $a_{H^+(s)}$, may be expressed in terms of the electrical potential at the surface, ψ_o (House 1994):

$$a_{H^+(s)} = a_{H^+} \exp\left[-e\,\Psi_o/(k_B T)\right], \qquad (10)$$

where k_B is the Boltzmann constant, T is temperature, a_{H^+}, $a_{H^+(s)}$ the activity of hydrogen ions in the bulk solution and at the surface respectively and e is the elementary charge. Typically, the surface charge of the clay leads to surface potentials < -300 mV and so substantial differences in the pH at the surface or interlayer and the bulk of the solution are possible. For example, if $\psi_o = -200$ mV, the difference between the bulk solution pH and the surface pH (pH(s)), ΔpH, is predicted from Eq. (10) as 3.4. This difference may explain why weakly basic compounds such as atrazine ($pK_a = 1.68$) are adsorbed strongly on H-montmorillonite at pH = 3.4, well above its pK_A (Bailey et al. 1968). Atrazine will be 10% associated at pH = 2.68 and even less at pH = 3.35. However, if $\Delta pH > 1.7$ at least 50% of the atrazine will be protonated at pH = 3.4 and able to enter cation exchange reactions at the surface.

It is interesting at this point to list some of the more important mechanisms that have been postulated to describe the molecular interactions with clays. These divide into the groups often described as: (1) weak van der Waals' or dispersion interactions caused by the interaction of fluctuating dipoles arising from the motions of the outer electrons of the adsorbate and the surface (2) electrostatic interactions, either charge attraction (cation-anion interaction), charge-dipole, dipole-dipole (and higher polar interactions) or induced dipole-charge/dipole interactions; (3) hydrogen-bonding; and (4) hydrophobic interaction when the intrinsic interactions of the water molecules dominate the energetics of the sorption process. Some examples of these reactions are noted as follows where R is a functional group, M^{z+} a cation with charge z and B represents a base

1. Hydrogen bonding with the siloxane surface:
   R-N-H ... O-clay
   R-C-H ... O-clay
   R-COOH ... O-clay
2. Hydrogen bonding with hydroxyl groups at the edge sites:
   R-C-O ... HO-clay
3. cation association:
   =C=O ... M^{z+}-clay
4. water bridging via polarised cation:

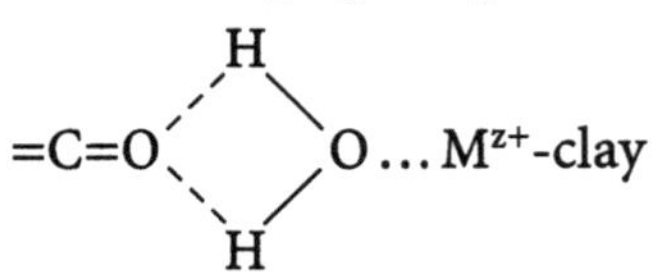

5. electrostatic:
   B + H^+-clay = HB^+-clay
   B + $M^{z+}(H_2O)$-clay = HB^+-clay + M^{z+}-OH-clay
   R-COO^- + H^+-clay = R-COOH-clay

4.3.1
Interactions of Basis Compounds

The exchange of organic bases with montmorillonite has been known for a long time. The reactions have been studied using concentrated solutions of the organic salt in water or by neutralising acid montmorillonite with the appropriate base. In this way a range of bases have been studied including aromatic amines, e.g. benzidine and aniline, as well as fluorene and the purine bases, adenine and quanine (Hendricks 1941). Measurements of the interlayer spacing and base-exchange capacity have enabled some postulates to be made about the orientation of the sorbed bases within the layer spacing. The adsorption of basic compounds to clay minerals, such as montmorillonite, is particularly strong when the solution $pH < pK_A$ ($pK_A = pK_w - pK_B$). In general, the affinity for the clay depends on the pH, temperature, the cations associated with the clay and the molecular structure of the compound. Often the separation of these contributing factors is difficult, and in some instances the role of impurities such as Al-Fe oxy/hydroxides, often associated with clay particles, also needs to be assessed. Other complicating factors include the dissolution and interaction of Al at the clay surface at low pH (so-called aluminisation of the clay), and the magnitude of the surface charge density (SCD) of the specific clay, i.e. montmorillonite clays from different origins and histories often possess different extents of isomorphous substitution and hence SCD as well as specific surface area. Clays with isomorphous substitution have a permanent negative surface charge which is essentially independent of the solution pH; however, as explained above, the charge on the edge sites is dependent on the solution pH. Hence, protonated basic compounds or cationic micro-organic compounds are capable of ion-exchange reactions with surface species, either through the formation of inner-sphere complexes or through polarised water acting as bridging links to the surface.

Table 4.2. Structures and properties of some s-triazines. pK_A is the acid dissociation constant of the protonated base and S is the aqueous solubility at 25 °C and pH = 7

Compound	Group on ring			pK_A	$S/10^{-4}$ (mol dm^{-3})
	2 Position	4 Position	6 Position		
Atrazine	Cl	NHC_2H_5	NHC_3H_7	1.68	1.6
Ametryn	SCH_3	NHC_2H_5	NHC_3H_7	4.00	8.6
Prometryn	SCH_3	NHC_3H_7	NHC_3H_7	4.05	1.7
Desmetryn	SCH_3	$NHCH_3$	NHC_3H_7	4.00	24.0

Probably the most studied compounds in this group are the s-triazine herbicides including the most important members, atrazine and simazine, but also including others such as ametryn, desmetryn, prometryn, dyrene, atratone and prometone. Examples of the structures are given in Table 4.2, together with the pK_As for the protonated molecules. Sorption studies with Na-montmorillonite have concluded that triazine sorption is generally a maximum in the vicinity of $pH = pK_A$ (Weber 1970). This indicates that the interaction of the positively charged triazine molecule with the negatively charged siloxane surface is a predominant mechanism at this pH. The detailed mechanism of the interaction is however still uncertain and is probably some combination of the following:

1. Direct ion exchange involving the charged triazine molecule and adsorbed cations on the negatively charged clay surface.
2. Protonation of the neutral molecule in the surface layer caused by the increased surface acidity, compared with the bulk solution, and hence greater polarisation of the water molecules at the surface and in the case of expandable clays, in the interlayer spacing.
3. In low pH solutions, surface aluminisation may lead to hydrogen bonds between the triazine (here labelled 'T') and polarised water molecules in the hydration sphere of the sorbed Al^{3+}. In these circumstances, ion exchange is not important and surface reactions such as:

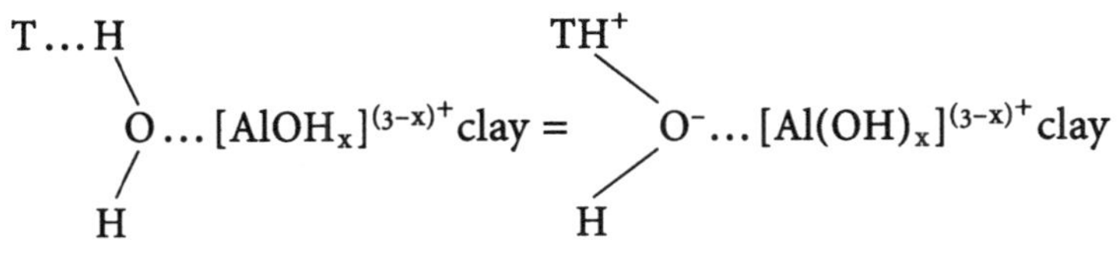

have been postulated (Calvet and Terce, 1975).

There is evidence for the penetration of triazines into the interlayer spacing of expandable clays (Cruz et al. 1968; Russell et al. 1968; Fruhstorfer et al. 1993). This is based on indirect evidence from measurements of the slow kinetics of desorption from Na-montmorillonites and the much higher adsorption capacity of Na-montmorillonite compared with illite and kaolinite (Gilchrist et al. 1993). Table 4.3 lists the adsorption capacities of atrazine on montmorillonite, kaolinite and illite, together with various surface characteristics taken from Gilchrist et al. (1993). The parameter n_c in the table is the labile surface sorption capacity in μmol g^{-1}, i.e. when $n_a = n_c$, the surface is completely occupied. There is a fundamental difference between Na and Ca montmorillonites, with the Na-clay possessing a greater affinity and sorption capacity for atrazine compared to the

Table 4.3. Comparison of the surface properties and sorption parameters for the sorption of atrazine on clays at pH = 5.1 ± 0.2 Gilchrist et al. (1993). SSA is the specific surface area in $m^2 g^{-1}$, cec the cation exchange capacity, K_d the distribution coefficient, SCD is the surface charge density (= ced/SSA) and n_c is the labile sorption capacity (equivalent is abbreviated val)

Clay	SSA ($m^2 g^{-1}$)	cec (mval g^{-1})	SCD (μval m^{-2})	n_c (μmol g^{-1})	n_c/SSA (nmol m^{-2})	K_d ($dm^3 kg^{-1}$)
Na-mont	727	1.10	1.51	2.28	3.14	7.60
Ca-mont	733	1.10	1.50	1.00	1.36	0.92
Kaolinite	84	0.18	2.14	0.32	3.81	0.04
Illite	219	0.36	1.64	0.47	2.15	0.17

Ca-clay. This is because the highly polarized Ca^{2+} ions in the interlayer are more ordered, with a strongly bound hydration-sphere impeding the entry of atrazine into interlamellar spacing. Other evidence for the penetration of triazines onto the interlayer spaces comes from shifts in the infrared NH stretching bands and X-ray diffraction data (Hermosin et al. 1982). The IR bands from 3400–3340 cm^{-1} in the prometryn-montmorillonite complex correspond to the protonated complex and compare to the band at 3240 cm^{-1} for the unprotonated prometryn or prometryn-montmorillonite system after displacement of the triazine with NH_4^+. The x-ray diffraction results indicate an interlayer spacing of 1.26 nm, corresponding to a single layer of prometryn in the interlamellar spacing with the triazine ring parallel to the siloxane layers. After displacement of the triazine by NH_4^+, the interlayer spacing reduces to 1.05 nm which is close to that measured for NH_4-montmorillonite and smaller than that for Ca-montmorillonite (0.96 nm). It is interesting to note that although the sorbed triazines are effectively non-available to the biota, they may become bioavailable on displacement from the clay, i.e. they may be released as the active compound rather than an inactive hydroxytriazine (Hermosin et al. 1982). It is also possible that the lower pH at the clay surface leads to greater chemical hydrolysis of the triazines (Skipper and Volk 1972), although other studies have indicated the lack of hydrolysis of atrazine by clay fractions (Goswami and Green 1971).

When the pH of the solution is greater than the pK_A of the compound, sorption of the neutral triazine molecule occurs on the siloxane surface. The affinity to the surface depends on the surface charge density. For example, smectites of low charge density have a greater affinity for atrazine than higher charge-density materials, indicating that the adsorbate is neutral rather than protonated (Laird et al. 1992). It is likely that surfaces with a high charge-density have sorbed layers of water and cations, with highly ordered structures, making it difficult for the neutral molecules to approach the surface. The actual mechanism of the interaction is still uncertain, although a hydrogen-bonding mechanism is most likely. In general, triazines may act as both hydrogen-bond acceptors and donors.

The kinetics of the interaction are particularly interesting. Gilchrist et al. (1993) have separated the kinetics according to: (1) labile adsorption on the surface with the adsorbate loosely bound, and (2) bound residues which are strongly sorbed and not easily released. Their results show that "labile" adsorption occurs on montmorillonite, kaolinite and illite and is generally complete in < 2 min (except with Na-montmorillonite which gave a slow phase as well). The reaction involving the bound residue occurred for all the clays except Ca-montmorillonite, and generally consisted of both fast and slow phases. The two-stage kinetics were attributed to a fast stage involving the uptake of the labile fraction on the surface; this led to a slower kinetic stage as the molecules diffused into the particles (intra-particle diffusion). As both kaolinite and illite gave similar behaviour to Na-montmorillonite, it is assumed that the slow kinetics associated with the bound residue, are not solely caused by movement into the interlayer spacing. It is probable that the slow kinetics of the labile fraction for Na-montmorillonite (but not Ca-montmorillonite) are associated with diffusion of atrazine into the interlayer spacing.

Another basic compound, 3-aminotriazole, with a $pK_B = 9.83$, has been shown to interact with montmorillonite by penetration into the lamellar spacing. The evidence for this comes from changes in the basal spacing of the clay and the appearance of C=N stretching of the exocyclic $C{=}NH_2$ in the polarised form, following sorption, i.e.:

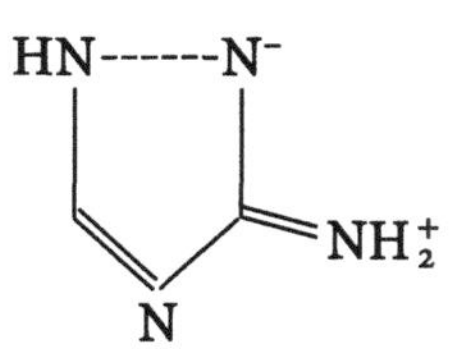

The normal frequency of the $C\text{-}NH_2$ bond as a molecule is 1630 cm^{-1} compared with 1690 cm^{-1} for the $C{=}NH_2^+$ bond.

It has been postulated (Morillo et al. 1991) that three mechanisms of interaction operate. The relative importance of these depends on the cationic form of the clay and the general solution conditions such as the pH. The mechanisms are:

1. Cation exchange through the formation of the protonated form:

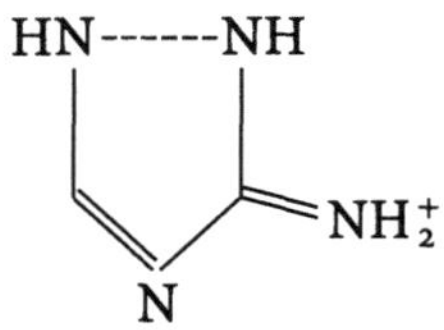

The protonation at a bulk solution pH value greater than the pK_A of the compound occurs because of the greater acidity in the interlamellar spacing – see Eq. (10) – and also because of the polarising power of the exchangeable cations, Me^{2+}

$$[Me(H_2O)_{x+1}]^{2+} + H_2O = [Me(OH)(H_2O)_x]^+ + H_3O^+,$$

leading to:

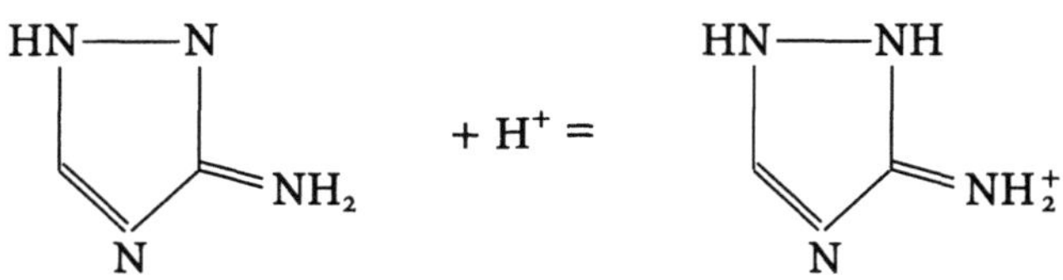

The cation then reacts through ion-exchange with the interlayer charge.

2. Charge-dipole interactions directly with the exchangeable cations

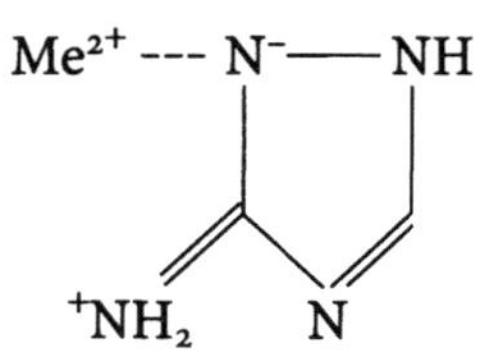

3. Water-bridging mechanism

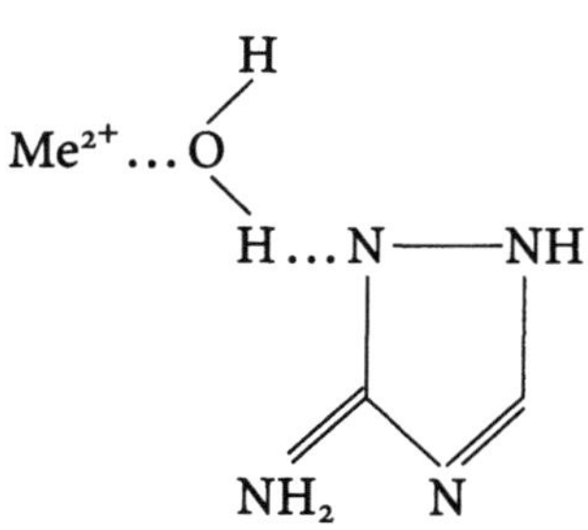

(neutral form)

The degree of polarisation of the aminotriazole by the cation determines the relative significance of mechanisms (1), (2) or (3). Hence, with Fe^{3+}-montmorillonite at pH 6.3, cation exchange predominates because of the high polarising power of the cation and leads to little release of the aminotriazole after washing the clay with water. In contrast, Na and Li-montmorillonite, at the same pH, produce an increase in basal spacing after contact with the herbicide. When the triazole is subsequently released by washing the clay, the lattice spacing reverts to its original value. Cations of intermediate polarisation, however, produce a decrease in the lattice spacing when the triazole is sorbed, implying a loss of water associated with the interlayer cations. The spacing reverts to its original value after the adsorbate is released. Again, this suggests a relative weak interaction of the aminotriazole, probably through a dipole-charge interaction (3. above) with the dipole replacing an inner-sphere water of co-ordination. For all the clays, the herbicide was not released from the interlayer space after washing with distilled water supporting the formation of a cationic form. If the pH is reduced to approximately 4, close to the pK_A of the triazole, the interaction is via the aminotriazolium cation, and the adsorbate is not displaced by washing the clay with water (Morillo et al. 1991), but undergoes normal exchange reactions with other cations. This is also evident for Al-montmorillonite (Russell et al. 1968).

Mechanism (1) is generally considered to be an important influence on base-clay interactions depending on the electric field strength of the inorganic cation, the basicity of the adsorbate, the water content of the system and the site of the layer charge. Strong bases such as NH_3 (pK_B = 4.7) are protonated by weakly polarising exchangeable cations on montmorillonite such as Na^+. Pyridine (pK_B = 8.8) is less so, whereas a weak base such as urea (pK_B = 13.8) is not protonated by exchangeable K^+, Na^+, Li^+, Ca^{2+} or Mg^{2+} in the clay. Protonation of urea only occurs in the presence of more strongly polarising cations such as Al^{3+}, Fe^{3+} or H^+ (for a review of the data see Rausell-Colom and Serretosa 1978). The degree of polarisation of the water of hydration also depends on the extent of the hydration, i.e. the water content of the clay. As the degree of hydration decreases, the polarisation and acidity of the interlayer water increases. For example Farmer and Mortland (1966), observed that pydrine molecules adsorbed on hydrated Mg-montmorillonite were not protonated until the clay was dehydrated.

A comparison of the sorption behaviour of several amino acids, at concentrations of about 10 μM on kaolinite (pH = 6.9) and montmorillonite (pH = 8.4), has also demonstrated the enhanced uptake of the basic amino acids compared with the neutral or acidic molecules. For example, montmorillonite (predominantly in the Na form) showed a much greater affinity for lysine and arginine compared with the neutral (e.g. alanine or valine) or acidic molecules (e.g. aspartic acid) (Hedges and Hare 1987). Pyridine and its derivatives also comprise an important class of environmental contaminants (see for example the review of Sims and O'Loughlin 1989). They are generally adsorbed through ionic mechanisms at $pH < pK_a$ as discussed above. It is known that compounds such as pyridine ($pK_a = 5.25$) and quinoline ($pK_a = 4.9$) compete for anionic sites on sediments at low pH when they are partially ionised but, at high pH, the neutral molecules adsorb independently, probably because of the dominance of the hydrophobic mechanisms (Zachara et al. 1987).

The phenylureas are also a group of weakly basic herbicides (e.g. fenuron and monuron have pK_as of -1 and -2), which adsorb to clays, with the affinity depending on the exchangeable cation but essentially independent of the pH. Again, the sorption has been found to depend on the polarising power of the exchangeable cation (Van Bladel and Moreale 1974) so that H-montmorillonite possesses a higher affinity for phenyl ureas compared to the Na or Ca-montmorillonites. In fact the K_d for monuron on montmorillonite at 3.5 °C has been shown to change in proportion to the electronegativity of the exchangeable cation in the order $H^+ > Al^{3+} > Mg^{2+} > Ca^{2+} > Na^+$ which is the same as the free energies of hydration of the cations. Hence an adsorption process based on a water-bridging mechanism (see 3. above) is very likely. There is also evidence that the adsorption of linuron on montmorillonite involves the interaction of the carbonyl group of the herbicide with the exchangeable cation of the clay either directly or through water bridging (Khan 1974).

Paraquat (1,1-dimethyl-4,4-bipyridinium di-cation) and diquat (1,1-ethylene-2,2-bipyridylium bromide) are bipyridinium herbicides which strongly associate with clay minerals. The sorption capacity of diquat is in the order: montmorillonite > vermiculite > illite > kaolinite. In expandable clays, the quats are incorporated into the interlayer spacing where they are resistant to degradation. The isotherms are typically Langmuirian when extended to high equilibrium concentration, e.g. > 50 mg dm^{-3}, with the sorption amount at the plateau essentially independent of the cationic form of the clay for Na^+, Ca^{2+}, and NH_4^+. At high salt concentrations, the quats are generally displaced in the order of efficiency: $NH_4^+ > K^+ > Ca^{2+} > Mg^{2+} > Na^+$ for kaolinite, illite and montmorillonite. For the paraquat/Ca-montmorillonite system, the interlayer spacing increases to a maximum 1.26 nm at the cation-exchange capacity of the clay. This is consistent with the co-planar bipyridylium ions lying flat within the interlamellar spacing of the clay lattice (Knight and Denny 1970).

Cationic surfactants, especially the dialkyl dimethyl ammonium compounds, also adsorb strongly to clay surfaces. The most widely occurring compound in this group is dihydrogenated tallow dimethyl ammonium chloride (DHTDMAC) used in fabric softeners:

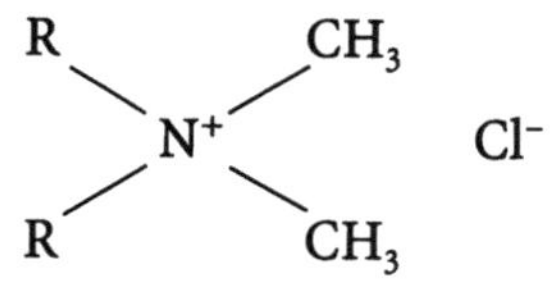

where R is the alkyl chain: C_{12} (maximum of 2%), C_{14} (1–5%), C_{16} (25–25%), C_{18} (60–70%) and C_{20} (maximum of 2%) with an average molecular weight in the range 567–573. The distribution coefficients, K_d, Eq. (7), for sediments are generally high, in the range 3000 to 85000 l kg^{-1} and desorption is usually extremely slow, so that for all practical purposes adsorbed DHTDMAC can be considered permanently bound to particles (ECETOC 1993). The adsorption of dimethyl ethyl octadecenyl ammonium bromide on kaolinite, illite and montmorillonite has demonstrated a high affinity for clays, with the sorption exceeding the base-exchange capacity (Grim et al. 1947). This implies a hydrophobic adsorption mechanism is also operative, perhaps similar to that found for anionic surfactants on kaolinite (House and Farr 1989). Certainly at high concentrations of surfactant, but below the critical micelle concentration (cmc) of the surfactant, bilayer structures occur so that the trimethylammonium head groups interact with the mineral surface, with a second layer stacked so that the head groups are exposed to the solution, e.g. hexadecyltrimethylammonium bromide on amorphous silica (Rennie et al. 1990). At low adsorbate concentrations, sorption by the formation of small surface aggregates of the surfactant molecule is likely (see, for example, Rupprecht and Gu 1991 for a detailed discussion). There is also evidence for the penetration of surfactant molecules in the interlayer spacing of montmorillonite to form a bilayer (Peker et al. 1995). Such hydrophobic domains provide sites for co-adsorption as demonstrated by Monticone and Treiner (1995). They demonstrated the adsorption of a hydrophobic solute, 3-phenoxyl-1-propanol, on adsorbed aggregates of cetylpyridium chloride at the silica surface. The co-adsorption was to be a function of the surface density of the cationic surfactant and hydrophobicity of the adsorbate and to be independent of the solution pH and ionic strength.

4.3.2
Interactions of Acidic Compounds

In general, the sorption of anionic compounds to clay is much lower than basic and cationic compounds under similar conditions of temperature and concentration. This is because of the inherent negative charge of the major clay surfaces at neutral pH. For example, the acidic amino acids, such as aspartic and glutamic acid, are weakly sorbed on Na-montmorillonite at pH = 8.4 (Hedges and Hare 1987). However, there is strong evidence that the sorption to kaolinite is controlled by the interaction of the anionics with the positively charged edge sites (pH < 6; see Sect. 4.1.3). This has been demonstrated for the interaction of benzoic acid ($pK_a = 4.2$) with a kaolinite sample (PZC = 5), when the sorption increases between pH 3 to 5 because of the increasing ionisation of the acid and occurrence of positive edge sites, and then decreases above pH 5 because of the increase in negative edge sites (Clarke 1972). Similar behaviour has also been found for aspartic and glutamic acids at pH = 6.9 sorbed to kaolinite, with the higher charge density of aspartic acid producing the greatest adsorption (Hedges and Hare 1987).

Experimental studies have also shown the sorption of anionic surfactants to clays, even in solution conditions where the clay surfaces are negatively charged as a result of the ionisation of SiOH and AlOH edge sites, e.g. edge sites of kaolinite (Poirier and Cases 1991). This may be a result of the interaction of the surfactant with inner-sphere cation complexes (in the O-plane, Fig. 4.2). The sorption of anionic surfactants on kaolinite, illite and montmorillonite is dependent on the alkyl chain length, ionic strength of the solution, pH and the concentration of competing anionic compounds in solution. In particular, the presence of $AlCl_3$ at pH values between 4 and 7 increases the sorption to kaolinite (Waymann 1963). The dependence of the sorption on alkyl chain length has also been found for the linear alkylbenzenesulphonates (LAS) on K-kaolinite at 25 °C, 0.01 M $KHCO_3$ at a pH of 7.6 to 7.8 (House and Farr 1989), and linear alkylsulphates under the same conditions but on a natural sediment containing clay minerals (Marchesi et al. 1991a, b). For the sorption of C_{10} to C_{13} LAS homologues, the values of the distribution coefficient, K_d, Eq. (7), increase as shown in Fig. 4.3. The distribution coefficient may be expressed in the form:

$$K_d = A \exp[-(\Delta G^0_R + i\Delta G^0_c)/RT], \quad (11)$$

where A is a constant, R is the gas constant, T the temperature and i refers to the alkyl chain length, i.e. i = 12 refers to C_{12}LAS. The term ΔG^0_c is the free-energy of adsorption associated with each-CH_2 in the alkyl chain

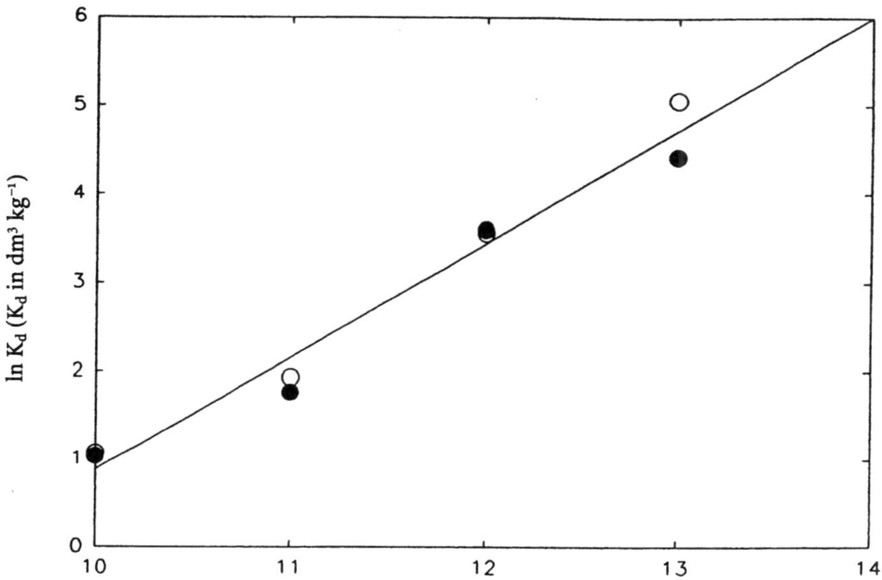

Fig. 4.3. Illustration of the application of Eq. (11) for C_{10} to C_{13} linear alkylbenzene sulphonates (LAS). An increase in the alkyl chain length leads to an increase in the adsorption affinity and distribution coefficient, K_d

and the residue, ΔG_R^0, accounts for the remaining adsorption energy. A linear regression of ln K_d with i (the alkyl chain length) leads to a value of $\Delta G_c^0 = 3.17 \pm 0.23$ kJ mol^{-1}, in reasonable agreement with the cohesive energy associated with the lateral interation of CH_2 groups evaluated by other investigators for different surfactants, and confirms the importance of the hydrophobic nature of the sorption interaction in these systems.

4.4
Case Study: Role of Clays in Complex Sediments

So far, the discussion has concentrated on examples of sorption mechanisms operating in reasonably well-defined adsorbate/clay systems in controlled solution conditions. This information is essential to our understanding of the mechanisms which control the partition of micro-organic compounds with clays in environmental conditions. In most situations however, sediments and soils are complex mixtures of in-organic and organic materials as described in Section 4.1.1 above. The role of individual components in influencing the sorption of particular micro-organic compounds in these matrices becomes difficult to determine. In response to this problem, the most widely adopted approach has been to use the organic matter or carbon composition of the soil/sediment to predict the sorption capacity. As noted by House et al. (1992) and Grundl and Small (1993), this approach may have serious limitations when colloids and clay components are present. The following section describes a case study to illustrate some of the complexities involved in understanding the interactions of selected compounds with river sediments.

The study site was the R. Windrush catchment in Oxfordshire, UK. The river starts in Inferior and Great Oolite White Limestone where the upper sampling site (site 1) was chosen (NGR SP 129228) and flows into Lias clay. A tributary, the R. Dikler, joins the R. Windrush in this section. The R. Dikler, although starting in a limestone area, mainly flows through an area dominated by Lower Lias clay. The second sampling site was on the R. Dikler near its confluence with the R. Windrush (site 2 at NGR SP 179223). The third site was chosen on the R. Windrush at a downstream location (site 3 at NGR SP 205131). The sites thus included areas of the catchment dominated by carbonates (site 1), clay (site 2) and a downstream site receiving water and sediment from both geologically distinct areas.

Details of the field sampling and analytical methods employed have been reported previously (House et al. 1992). In brief, the fieldwork involved:

1. Sampling water for major-ion analysis and pesticide analysis at each site.
2. Sampling bed sediments (ca to a depth of < 5 cm) and sieving through a 1-mm sieve; fine particulate matter in suspension was stored in separate containers and later centrifuged and combined with the rest of the sediment.
3. Water samples containing suspended sediments were collected in glass containers (about 32 l) for separation and analysis of the solids. The suspended sediment concentration was also measured.
4. On-site measurement of dissolved oxygen, pH, temperature and conductivity as well as estimates of the water discharge were made at each site.
5. Samples of the source sediments, Lower Lias clay and Great Oolite White Limestone were also collected for laboratory studies.

At the laboratory, the samples of water, bed and suspended sediment were analysed for pesticides (a range of compounds but including atrazine, simazine, prometryn, malathion, parathion and fenitrothion). The mineral composition of the sediments and source minerals was determined by X-ray diffraction and particle-size fractionation by sedimentation. The organic carbon and specific surface area were also measured (House et al. 1992). In addition, batch sorption experiments with the bed-sediment from the three sites and the source minerals, enabled K_d values to be calculated for simazine, atrazine, prometryn, malathion, parathion and fenitrothion.

The water chemistry at the three sites at the time of the visit was very similar. The river water was hard ($[Ca^{2+}]$ = 2.4–2.7 mM) with pH at sites 1 and 3 in the range 8.2–8.4. The R. Dikler site had a lower pH = 7.7, with slightly elevated concentrations of dissolved phosphate (2.9 μM) and nitrate (0.98 mM) compared with the other sites. The dissolved oxygen concentrations were between 79 and 105% saturation and water temperature between 6.8 and 7.3 °C. The suspended sediment concentrations were 5.1, 26.4 and 13.7 mg dm^{-3} for sites 1–3 respectively. Hence, the water composition at the sites was sufficiently similar that differences in composition were not likely to be important in determining the adsorption of the pesticides.

The results of the analysis of pesticides in the water showed the presence of the triazines as shown in Table 4.4. The triazines were not detected in the bed or suspended solids at any site. Only parathion was detected in the bed and suspended sediments with concentrations in the suspended sediment greater than the bed material. Some organochlorine pesticides were also detected in the sediments and to a much lesser extent in the water. Based on these results, the sorption behaviour of these compounds on the bed sediments and source minerals was investigated in

Table 4.4. Concentrations of triazines found in whole water samples (aq) and parathion in sediments. Units of concentration are ng dm^{-3} in aqueous and mg kg^{-1} (dry mass) in sediments. The attachment of C indicates that the quantification was by GC/NPD and confirmation by GC/MS

Pesticide	Site 1	Site 2	Site 3
Simazine (aq)	4.9	1.5 C	22 C
Atrazine (aq)	15 C	7 C	17 C
Prometryn (aq)	ND	ND	1.2 C
Parathion (bed)	0.3 C	0.6 C	1.0 C
Parathion (suspended)	13 C	3.3 C	8.8 C

ND, not detected

Table 4.5. Distribution coefficients, K_d [$dm^3\ kg^{-1}$; see Eq. (7)], for 1-mm sieved bed sediments and source minerals from the R. Windrush measured in the dark at 10 ± 0.2 °C in 0.01 M $CaCl_2$ solution. The solution composition varied only slightly during the measurements, i.e. pH 7.6–7.9 and dissolved oxygen of 79 to 97% saturation during the 24-h contact between the pesticide and adsorbent. The method involved the addition of freeze-dried sediments to 220 ml of 0.01 M $CaCl_2$ containing 3–30 mg dm^{-3} of each pesticide. The adsorbent concentrations were 0.03 to 0.14 kg dm^{-3}, depending on the adsorption affinity of the sediment, i.e. the sediment concentration was adjusted to produce a decrease of approximately 50% in the pesticide concentration after adsorption

Adsorbent	Triazines			Fenitrothion	Malathion	Parathion
	Sim′	Atr′	Prom′			
Site 1	7	7	1	37	38	38
Site 2	6	6	7	103	208	108
Site 3	33	41	38	202	81	301
White limestone	3	4	0	1	1	2
Lower Lias clay	4	3	10	164	118	161

more detail. The distribution coefficients, Eq. (7), K_d, calculated from the change in pesticide concentration following 24 h contact with the solution of the pesticide in 2 mM $CaCl_2$, pH 7.6–7.9, are shown in Table 4.5. The triazines exhibit a low affinity for all the bed sediments and Lower Lias clay when compared with fenitrothion, malathion and parathion. The adsorption of all the compounds to the Great Oolite White Limestone is low, with slightly greater adsorption of simazine and atrazine in comparison to the other compounds. If the distribution coefficients are normalised with respect to the content of organic carbon, OC, i.e. $K_{oc} = 100\ xK_d$/(percentage by mass of OC), only the sediment from site 3 gives a good correlation with the octanol-water partition coefficients of the pesticides, i.e. $\log K_{oc} = 1.34 (\pm 0.26) + 0.66 (\pm 0.09) \log K_{ow}$, where the

Fig. 4.4. Comparison of the composition and specific surface areas (*SSAs*) of the bed (< 1 mm size) and suspended sediments studied. The organic carbon (*OC*) is expressed as percentage of mass of the dry weight of the sediment, the clay fraction is the < 2 μm fraction measured by sedimentation and the specific surface area, SSA/m^2g^{-1}, is from nitrogen adsorption and analysis by the BET method. The results for quartz (*Q*) and calcite (*Cal*) are the percentage composition from the whole rock analysis by X-ray diffraction, whereas the results for kaolinite (*Kaol*), illite and expandable clays (*Exp*) are from the X-ray diffraction of the clay fraction

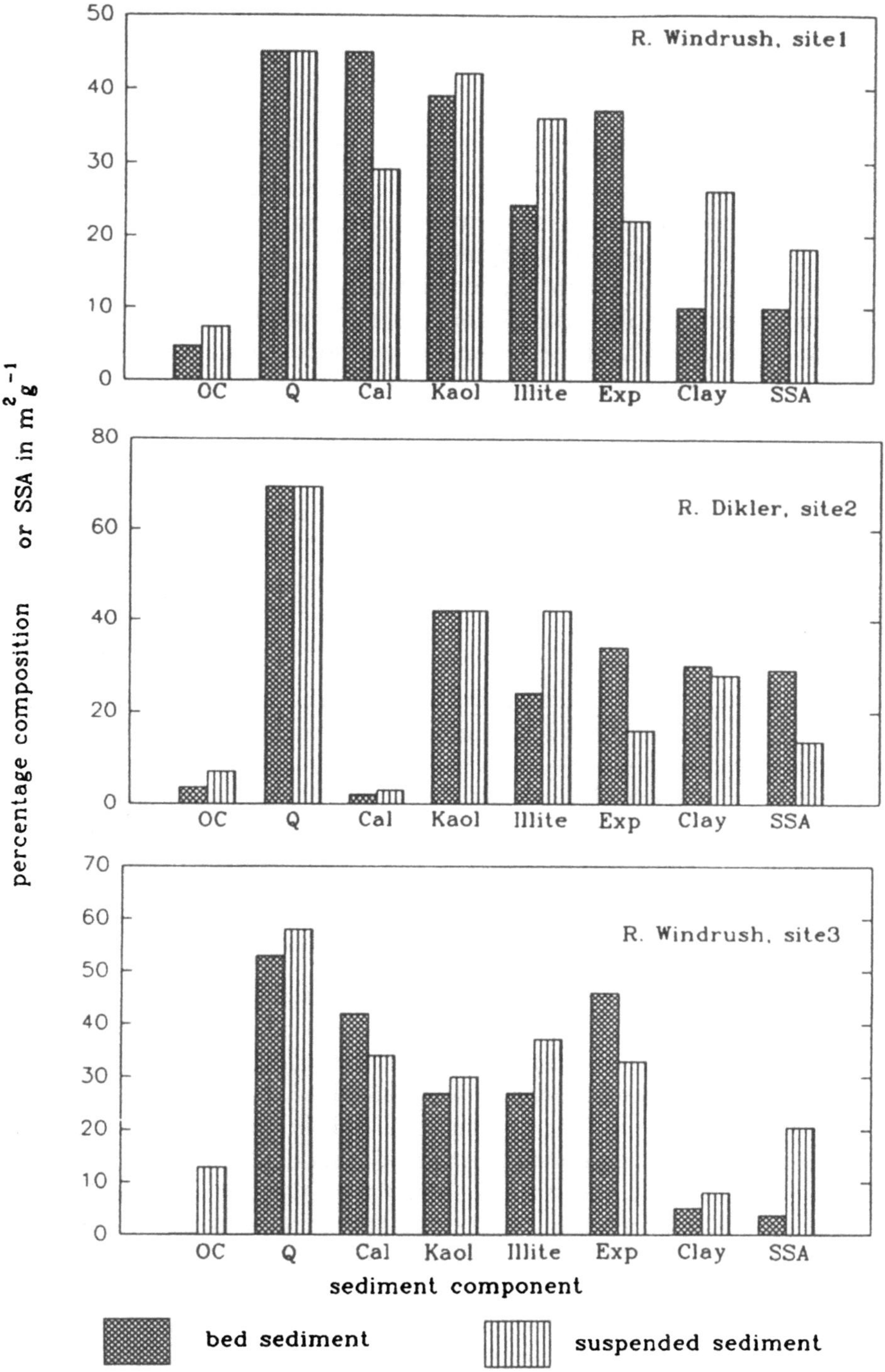
R. Windrush, site1
R. Dikler, site2
R. Windrush, site3
percentage composition or SSA in m² g⁻¹
OC
Q
Cal
Kaol
Illite
Exp
Clay
SSA
sediment component
bed sediment
suspended sediment

standard deviations are shown in brackets. However, a good correlation is also obtained for the distribution coefficient normalised with respect to the specific surface area of the sediment, i.e.:

$$\log K_d/SSA = -1.00\,(\pm 0.27) + 0.66\,(\pm 0.09)\,\log K_{ow}.$$

This suggests that the interaction of the pesticides at this site is by a hydrophobic interaction with a surface film on the particulates rather than with discrete organic particles.

Further understanding of the sorption behaviour of this group of compounds is only possible with a more detailed knowledge of the mineralogy and composition of the sediments and source minerals. The results of the X-ray diffraction studies, organic carbon and specific surface area measurements are shown in Fig. 4.4 for the bed and suspended sediments and Fig. 4.5 for the source minerals. The organic carbon contents are expressed as percentage by mass, the quartz and calcite components are percentage composition by whole rock analysis and the clay fractions (kaolinite, illite, expandable clays) are from a separate analysis of the <2 µm size fraction. The mass percent <2 µm determined by sedimentation, is labelled the clay fraction. The key features of these results are:

1. As expected, the two major source sediments are quite different in composition. The White Limestone is predominantly calcite with a low specific surface area, SSA, with the small clay component consisting of mainly expandable clays and illite. In contrast, the Lower Lias clay has a much higher SSA and is predominantly quartz with about equal proportions of clays and calcite. The clays are mainly expandable clays and kaolinite with no detectable illite component. The organic carbon composition of these minerals is generally lower than found for the sediments.
2. The composition of the bed and suspended sediments from the same site are very similar (see Fig. 4.4).
3. Site 1 had the highest proportion of calcite in the bed and suspended sediments and the suspended sediment was enriched with the clay component relative to the bed material. This is consistent with the low concentration of suspended sediment (5.1 mg dm^{-3}), found at this site.
4. Similarly, at site 3, the suspended sediment was enriched with clays and organic matter relative to the bed sediment which again is reflected in the higher SSA of the suspended material.
5. Site 2 showed different results with the SSA of the bed material higher than the suspended sediment and likewise a slight increase in the clay component in the bed sediment. The concentration of calcite in the

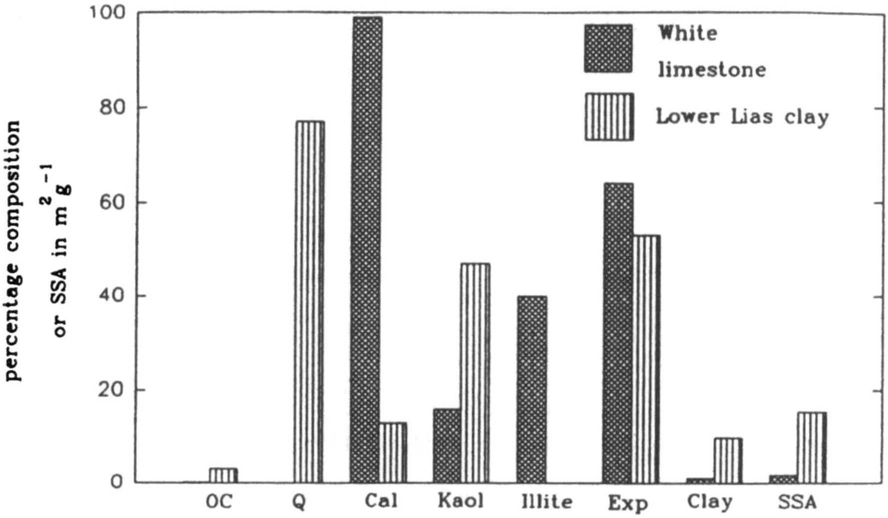

Fig. 4.5. Composition of the two major source minerals in the R. Windrush catchment, Great oolitic White limestone and Lower Lias clay. The measurements are the same as described in the legend to Fig. 4.4

sediments was low at this site which is in agreement with the composition of source minerals in this region of the catchment.

The results of a pairwise multiple regression analysis are shown in Table 4.6. The mineral composition was estimated from the percentage clay determined from the whole rock analysis and the detailed composition of the clay fractions with a correction for the organic matter content. This shows there is a reasonable correlation between the adsorption behaviour of the triazines and parathion and fenitrothion but not malathion. Parathion and fenitrothion show the strongest correlation with the organic carbon content of the adsorbent. The triazines show the worst correlation with both OC and SSA. Malathion gives the best correlation between the distribution coefficients and clay contents or specific surface areas. Basically, the clay content of sediment determines the specific surface area of the freeze-dried sediment. Attempts to identify the clay fraction responsible for the correlation were unsuccessful with equally good correlations with both the kaolinite and expandable clay components; the linear relationships are shown in Fig. 4.6. These results illustrate some of the complexities in predicting the adsorption of modern pesticides on natural materials. The fact that the traditional relationship between log K_d/OC and log K_{ow} is poor for the upstream sites (sites 1 and 2) suggests that

Table 4.6. Comparison of the correlation coefficients from a pairwise multiple regression analysis of the key parameters. Where the compound names are abbreviated, the first three letters of the name are used. Key: OC, organic carbon (% by mass); SSA, specific surface area by nitrogen adsorption and BET analysis in units of $m^2 g^{-1}$. Coefficients greater than 0.8 are in boldface type

Parameter	Sim	Atr	Pro	Fen	Mal	Par	OC
Atrazine	**0.998**	–	–	–	–	–	–
Prometryn	**0.955**	**0.948**	–	–	–	–	–
Fenitrothion	0.666	0.638	**0.841**	–	–	–	–
Malathion	– 0.016	– 0.048	0.153	0.487	–	–	–
Parathion	**0.848**	**0.829**	**0.959**	**0.959**	0.344	–	–
OC	0.634	0.612	**0.806**	**0.935**	0.680	**0.907**	–
Quartz	– 0.016	– 0.171	0.135	0.582	0.746	0.383	0.522
Calcite	– 0.017	0.033	– 0.174	– 0.593	– 0.885	– 0.408	0.621
Kaolinite	– 0.086	– 0.112	0.091	0.429	**0.990**	0.274	0.642
Illite	0.073	0.065	0.091	0.136	0.797	0.126	0.448
Expandable	– 0.139	– 0.168	0.077	0.484	**0.975**	0.293	0.649
SSA	– 0.070	– 0.096	0.103	0.436	**0.93**	0.285	0.648

other controls are important. This may apply to other rivers in upstream reaches where sewage inputs are low and minerals from surface or subsurface flow enter. Minerals from surface soils may already contain associated pesticides, bound by mechanisms discussed in Section 4.3. Depending on the mechanism of binding and whether the micro-organics have entered the interlayer spacing of expandable clays, the kinetics of the desorption may control the concentrations in the associated water. The results of this study also suggest that in situations where NOM controls the sorption behaviour once the particles are in the river, this organic material is likely to be associated with particle surfaces forming attached layer(s). Because of the relatively high specific surface area of clays, they are likely to have an important role in adsorption even when the micro-organic interacts by a hydrophobic mechanism.

4.5 Conclusions

There is no doubt that the interaction of micro-organic compounds with clay surfaces is potentially an important mechanism controlling their fate in surficial environments. The sorption affinity between clay surfaces and compounds in solution depends on the nature of the clay/water interface

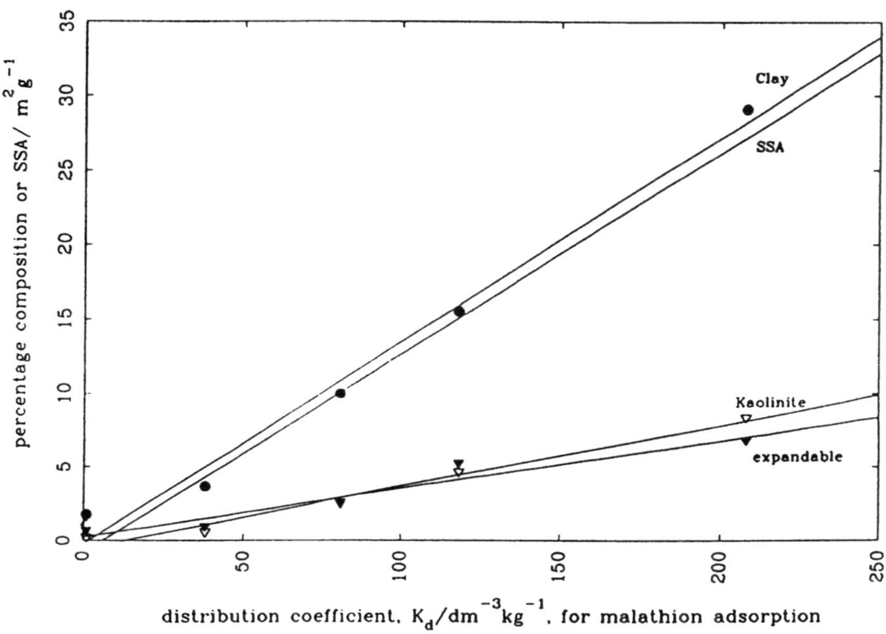

Fig. 4.6. Regression relationships between the distribution coefficient, K_d, for malathion adsorption at 10 ± 0.2 °C to the bed sediments and source minerals and the specific surface area of the adsorbents ($SSA/m^2 g^{-1}$), the percentage by mass of clay and the percentage composition of kaolinite and expandable clays in the minerals

and chemical properties of the individual compounds. In particular, the pH and concentration of exchangeable cations in solution are crucial factors controlling adsorption. Only specific adsorption mechanisms have been elucidated for a few systems. In these, hydrogen bonding, cation association between the clay and micro-organic compound, water bridging via polarised cations, electrostatic as well as hydrophobic mechanism have played a role. In the past, much emphasis has been placed on the role of natural organic matter in determining the adsorption on natural particles. However, clay surfaces may be more important than hitherto recognised because of their high specific surface area and strong interaction with a multitude of compounds. The more basic compounds are known to penetrate the interlayer spacing of expandable clays such as montmorillonite by a slow diffusion mechanism. This affects the kinetics of their release and degradation in soils and sediments.

Most of the research in this area has been concerned with the adsorption affinity of whole soils and sediments containing clays but few studies have attempted to relate adsorption/desorption measurements to individual mineral components or separate the contributions of natural organic matter from the clay components. There is also a poor link between laboratory and field studies and theoretical modelling of the sorption data, either for mineral/clay components or complex mixtures of clays and other natural matter occurring in sediments and soils. To date, the main emphasis has been on adsorption with few studies examining desorption processes. With recent advances in understanding the structure of the clay/mineral-water interface, it is now possible to examine the interactions in more detail to identify important mechanisms and provide a better understanding of adsorption and desorption and the kinetics describing these processes.

Acknowledgement. The author thanks the Natural Environment Research Council, UK, for financial support.

References

Bailey GW, White JL, Rothberg T (1968) Adsorption of organic herbicides by montmorillonite: role of pH and chemical character of adsorbate. Soil Sci Soc Am Proc 32:222–234
Bernstein F (1960) Distribution of water and electrolyte between homoionic clays and saturated NaCl solution. Clays Clay Miner 8:122–149
Calvet R (1989) Adsorption of organic chemicals in soils. Environ Health Perspect 83:145–177
Calvet R, Terce M (1975) Adsorption de l'atrazine par des montmorillonites-Al. Ann Agron 26:693–707
Charalambos P, Hayes KF (1966) Distinguishing between interlayer and external sorption sites of clay minerals using X-ray absorption spectroscopy. Colloids 107:89–96
Clarke CD (1972) Influence of pH on the adsorption of benzoic acid by kaolin. Pharm J 209:44–45
Costerton JW, Lewandowski Z, deBeer D, Caldwell D, Korber D, James G (1994) Biofilms, the customized microniche. J Bacteriol 176:2137–2142
Cruz M, White JL, Russell JD (1968) Montmorillonite-s-triazine interaction. Isr J Chem 6:315–323
ECETOC (1993) DHTDMAC: Aquatic and terrestrial hazard assessment. European centre for Ecotoxicology and Toxicology of Chemicals, Technical Repaort No 53, Brussels
Everett DH (1992) Basic principles in colloid science. Royal Society of Chemistry, London
Farmer VC, Mortland MM (1966) Infrared study of the coordination of pyridine and water to exchangeable cations in montmorillonite and saponite. J Chem Soc (A):344–351
Fox LE (1991) Phosphorus chemistry in the tidal Hudson river. Geochim Cosmochim Acta 55:1529–1538

Francioso O, Bak E, Rossi N, Sequi P (1992) Sorption of atrazine and trifluralin in relation to the physico-chemical characteristics of selected soils. Sci Total Environ 123/124:503–512

Fruhstorfer P, Schneider RJ, Weil L, Niessner R (1993) Factors influencing the adsorption of atrazine on montmorillonitic and kaolinitic clays. Sci Total Environ 138:317–328

Gilchrist GFR, Gamble DS, Kodama H, Khan SU (1993) Atrazine interactions with clay minerals: kinetics and equilibria of sorption. J Agric Food Chem 41:1748–1755

Goswami KP, Green RE (1971) Microbial degradation of the herbicide atrazine and its 2-hydroxy analog in submerged soils. Environ Sci Technol 5:426–429

Grim RE, Allaway WH, Cuthbert FL (1947) Reaction of different clay minerals with some organic cations. J Am Ceram Soc 30:137–142

Grundl T, Small G (1993) Mineral contributions to atrazine and alachlor sorption in soil mixtures of variable organic carbon and clay content. J Contam Hydrol 14:117–128

Hedges JI, Hare PE (1987) Amino acid adsorption by clay minerals in distilled water. Geochim Cosmochim Acta 51:255–259

Hendricks SB (1941) Base exchange of clay mineral montmorillonite for organic cations and its dependence upon adsorption due to van der Waals' forces. J Phys Chem Ithaca 45:65–81

Hermosin MC, Cornejo J, White JL, Hess FD (1982) Bioavailability of s-triazines adsorbed on montmorillonite. J Agric Food Chem 30:728–733

Hiemstra T, van Riemsdijk WH, Bruggenwert MGM (1987) Proton adsorption mechanism at the gibbsite and aluminium oxide solid/solution interface. Neth J Agric Sci 35:281–293

House WA (1994) The role of surface complexation in the dissolution kinetics of silica: effects of monovalent and divalent ions at 25 °C. J Colloid Interface Sci 163:379–390

House WA, Farr IS (1989) Adsorption of sulphonates from detergent mixtures on potassium kaolinite. Colloids Surf 40:167–180

House WA, Orr DR (1992) Investigation of the pH dependence of the kinetics of quartz dissolution at 25 °C. J Chem Soc Faraday Trans 88:233–241

House WA, Ou Z (1992) Determination of pesticides on suspended solids and sediments: investigation of the handling and separation. Chemosphere 24:819–832

House WA, Rae JE, Kimblin RT (1992) Source-sediment controls on the riverine transport of pesticides. Brighton Crop Protection Conference-Pests and Diseases, 1992, pp 865–870

Khan SU (1974) Adsorption of linuron by montmorillonite. Can J Soil Sci 54:235–237

Knight BAG, Denny PJ (1970) The interaction of paraquat with soil: adsorption by an expanding lattice clay mineral. Weed Res 10:40–48

Kookana RS, Aylmore AG, Gerritse RG (1992) Time-dependent sorption of pesticides during transport in soils. Soil Sci 154:214–225

Kuchler-Krischun J, Kleiner J (1990) Heterogenously nucleated calcite precipitation in Lake Constance. A short time resolution study. Aquat Sci 52:176–197

Kummert R, Stumm W (1980) Surface complexation of organic acids on hydrous g-Al_2O_3. J Colloid Interface Sci 75:373–385

Laird DA, Barriuso E, Dowdy RH, Koskinen WC (1992) Adsorption of atrazine on smectites. Soil Sci Soc Am J 56:62–67

Marchesi JR, House WA, White GF, Russell NJ (1991a) A comparative study of the adsorption of linear alkyl sulphates and alkylbenzene sulphonates on river sediment. Colloids Surf 53:63–78
Marchesi JR, Russell NJ, White GF, House W (1991b) Effects of surfactant adsorption and biodegradability on the distribution of bacteria between sediments and water in a freshwater microcosm. Appl Environ Microbiol 57:2507–2513
Monticone V, Treiner C (1995) Effect of pH and ionic strength on the adsorption of cetylpyridinium chloride and the co-adsorption of phenoxypropanol at a silica/water interface. Colloids Surf 104:285–293
Morillo E, Perez-Rodriguez JL, Maqueda C (1991) Mechanisms of the interaction between montmorillonite and 3-aminotriazole. Clay Miner 26:269–279
Newman ACD, Brown G (1987) The chemical constitution of clays. In: Newman ACD (ed) Chemistry of clays and clay minerals. Mineralogical Soc Mon No 6. Longman Harlow, New York, pp 1–129
Papelis C, Hayes KF (1996) Distinguishing between interlayer and external sorption sites of clay minerals using X-ray absorption spectroscopy. Colloids Surf 107: 89–96
Peker S, Yaper S, Besun N (1995) Sorption behaviour of a cationic surfactant on montmorillonite. Colloids Surf 104:249–257
Poirier JE, Cases JM (1991) Anionic surfactant adsorption onto silicate minerals. The role of cations. Colloids Surf 55:333–344
Rausell-Colom JA, Serretosa JM (1978) Reactions of clays with organic substances. In: Newman ACD (ed) Chemistry of clays and clay minerals. Mineralogical Soc Mon No 6. Longman Harlow, New York, pp 371–422
Rennie AR, Lee EM, Simister EA, Thomas RK (1990) Structure of cationic surfactant layer at the silica-water interface. Langmuir 6:1031–1034
Rupprecht H, Gu T (1991) Structure of adsorption of layers of ionic surfactants at the solid/liquid interface. Colloid Polymer Sci 269:506–522
Russell JD, Cruz MI, White JL (1968) The adsorption of 3-aminotriazole by montmorillonite. J Agric Food Chem 16:21–24
Schindler PW, Stumm W (1987) The surface chemistry of oxides, hydroxides and oxide minerals. In: Stumm W (ed) Aquatic surface chemistry. Wiley-Interscience, New York, pp 83–100
Sims GK, O'Loughlin EJ (1989) Degradation of pyridines in the environment. Crit Rev Environ Control 19:309–340
Skipper HD, Volk VV (1972) Biological and chemical degradation of atrazine in three Oregon soils. Weed Sci 20:344–347
Tipping E (1990) Interactions of organic acids with inorganic and organic surfaces. In: Perdue EM, Gjessing GT (eds) Organic acids in aquatic ecosystems. Wiley, New York, pp 209–221
Van Bladel R, Moreale A (1974) Adsorption of fenuron and monuron (substituted ureas) by two montmorillonite clays. Soil Sci Soc Am Proc 38:244–249
Velde B (1992) Introduction to clay minerals. Chapman and Hall, London
Wayman CH (1963) Surfactant sorption on heteroionic clay minerals. In: Rosenquist IT (ed) International clay conference proceedings. Macmillan, New York, 376 pp
Weber JB (1970) Adsorption of s-triazines by montmorillonite as a function of pH and molecular structure. Soil Sci Soc Am Proc 34:401–404
Wehrli B, Wieland E, Furrer G (1990) Chemical mechanisms in the dissolution kinetics of minerals; the aspects of active sites. Aquat Sci 52:1–31

Wieland E, Wehrli B, Stumm W (1988) The coordination chemistry of weathering. I. A generalisation on the dissolution rates of minerals. Geochim Cosmochim Acta 52:1969–1981

Wu S, Gschwend PM (1986) Sorption kinetics of hydrophobic organic compounds to natural sediments. Environ Sci Technol 20:717–7725

Zachara JM, Ainsworth CCD, Cowan CE, Thomas BL (1987) Sorption of binary mixtures of aromatic heterocyclic compounds on subsurface materials. Environ Sci Technol 21:397–405

Glossary

a_{H^+} hydrogen ion activity; (s) indicates the value at the surface plane
c_e equilibrium concentration of micro-organic compound in solution, Eq. (7)
c_{24} concentration of micro-organic compound after 24 h of contact with adsorbent
cmc critical micelle concentration
DHTDMAC dihydrogenated tallow dimethyl ammonium chloride
DG free-energy of adsorption associated with each CH_2 group
K_A acid dissociation constant
K_B base dissociation constant
K_d distribution coefficient (mass of adsorbate per unit mass of adsorbent/solution concentration of the adsorbate)
K_l Langmuir adsorption constant
K_f Freundlich adsorption constant
K_{oc} distribution coefficient normalised with respect to organic carbon
K_w ionic product of water
k_B Boltzmann constant
K_{ow} octanol-water partition coefficient
n Freundlich adsorption constant, Eq. (9)
n_a sorption amount per unit mass of adsorbent
n_m monolayer capacity of Langmuir isotherm, Eq. (8)
NOM natural organic matter
OC organic carbon content of soil or sediment (% by mass of dry weight)
OM organic matter content usually determined by combustion (% by mass of dry weight)
pK $pK = -\log(K)$
PZC point of zero charge of the mineral surface
SSA specific surface area of the particles in $m^2 g^{-1}$
w mass of adsorbent
ψ_o electrical potential at the surface

Chapter 5

The Biogeochemical and Ecological Significance of Interactions between Colloidal Minerals and Trace Elements

T. A. Jackson

5.1 Introduction

Heavy metals and metalloids have critically important biological effects, both beneficial and harmful. In trace quantities, many of these elements function as micronutrients, and certain ones (e.g. Cu and Zn) are essential to all living things; but they are poisonous when their concentrations in the environment or in organisms are too high, and some of the more toxic ones (e.g. Hg and Pb) are not known to perform any essential biochemical functions (Bowen 1966; Kieffer 1991). The most toxic elements are injurious or lethal to sensitive organisms even at very low ambient levels; for instance, an aqueous Cu^{2+} concentration as low as 0.01 µg/ml may kill green algae (Bowen 1966). Organisms accumulate heavy metals and metalloids from their surroundings to a greater or lesser extent (Bowen 1966) and may acquire undesirably high levels of such elements as a result of pollution or other conditions conducive to bio-accumulation (Hg in aquatic ecosystems being a particularly notorious example).

Trace amounts of heavy metals and metalloids from natural sources are ubiquitous in soils, aquatic environments, and the atmosphere. However, natural background levels vary from place to place owing to variations in the composition of bedrock and uneven distribution among different environmental compartments (e.g. preferential accumulation in fine sediments enriched in organic matter and sulfides) (Bowen 1966; Nriagu 1989; Painter et al. 1994; Rasmussen 1994); and temporal variation results from phenomena such as episodes of vulcanism (Nriagu 1989). Geographical areas of trace element deficiency are fairly common and are much more widespread than areas of excessively high levels, although in some regions bio-available forms of certain elements are naturally abundant enough to be toxic, as with Se in parts of the western United States (Schütte 1964; Bowen 1966). A deficiency may arise either from lack of an essential element in the environment (e.g. in local rocks and soils) or from limited bio-availability of the element, as with metals bound by humic acids in peaty soils and dystrophic lakes (Goodman and Cheshire 1973; Thornton 1979; Jackson and Hecky 1980).

Since the onset of industrialisation, pollution of air, water, and soil with heavy metals (e.g. Hg, Cd, and Pb) and metalloids (e.g. As) has hugely augmented natural background levels, creating a serious, extensive problem (Andren and Nriagu 1979; Nriagu and Pacyna 1988; Nriagu 1989; Lindqvist et al. 1991; Slemr and Langer 1992; Candelone et al. 1993). On a global scale, emissions due to human activity may rival or greatly exceed natural emissions, and in severely polluted regions (e.g. within several kilometers of a smelter, chlor-alkali plant, or other source of pollution) anthropogenic contamination is far in excess of input from natural sources. The pollution may be confined to a certain locality, as with Hg in Minamata Bay, Japan (D'Itri 1991), or spread over hundreds or thousands of kilometers, as in the fluvial transport of Hg from a point source (Fimreite and Reynolds 1973; Bishop and Neary 1976; Parks and Hamilton 1987; Parks et al. 1991, 1992) and the contamination of Arctic lands, Alpine peaks, and the open sea by atmospheric transport from distant industrial centres (Murozumi et al. 1969; Weiss et al. 1971; Lindqvist et al. 1991; Barrie et al. 1992; Lockhart et al. 1992; Slemr and Langer 1992; Candelone et al. 1993; Zechmeister 1995). The toxicity of such pollutants to organisms exposed to them ranges from chronic to acute, and from sublethal to deadly; and there have been adverse economic consequences, such as the closing down of fisheries owing to high methyl Hg levels in fish. Occasionally, pollution has led to disastrous mass poisoning (as with Hg in Minamata and Niigata, Japan). It is not known whether the attenuated pollution of remote areas by long-range atmospheric transport is having toxic effects on the native biota, but the occurrence of high metal concentrations in organisms inhabiting such regions (e.g. Arctic marine animals) is a cause for concern (Muir et al. 1992; Wagemann and Stewart 1994).

The bio-accumulation kinetics and biological effects of a heavy metal or metalloid depend on many variables: (1) the abundance of the element in the environment; (2) properties of the element (electron configurations, ionisation potentials, etc.) which control its chemical reactions (e.g. bonding by ligands and activation or inhibition of enzymes); (3) synergistic and antagonistic effects of other elements; (4) characteristics and activities of the organisms exposed to the element (mechanisms of detoxification, active and passive uptake mechanisms, physiological processes that regulate element levels, feeding habits, food chain relations, habitat selection, age, body size, growth rate, stage of development, biomass as related to "biodilution," etc.); and, of particular importance here, (5) the bio-availability of the element, which is determined by chemical speciation, sorption, complexing, and competition with other elements in the environment, and is controlled by a wide range of environmental and biological factors (e.g. pH, Eh, Cl^- ions, dissolved O_2, sulfides, humic matter, labile organic matter, colloidal minerals, primary productivity, and various microbial activities). Although major spatial variations in trace element abundance in the environment are reflected in the chemical composition of the biota (Fimreite and Reynolds 1973; Bishop and Neary 1976; Parks and Hamilton 1987; Parks et al. 1991, 1992; Muir et al. 1992), reactions between trace elements and organisms (e.g. bio-accumulation, toxic effects, and microbial methylation) are apt to correlate much more significantly with the abundance of bio-available forms of the element, and with factors that control bio-availability, than with the total abundance of the element in the environment (Jackson 1987, 1988a, 1991, 1993b; Jackson et al. 1993). Colloidal minerals (clay minerals and the oxides, hydroxides, and oxyhydroxides, i.e. "hydrated" or "hydrous" oxides, of Fe, Mn, and Al) are among the most important environmental factors that control the bio-availability, ecological effects, biogeochemical cycles, and distribution of trace elements in ecosystems. It is primarily with the effects of these minerals, as well as associated substances and environmental variables that affect their interactions with trace elements, that we will be concerned in this review.

At this point, clarification of terminology would be in order. In earth science usage, particles < 2 µm in diameter are classed as "clay-sized" or "colloidal" (Grim 1968), although chemists define the colloidal size range as $\sim$ 0.001–1 µm (Weast and Astle 1982). Heterogeneous unconsolidated material made up chiefly of colloidal minerals is designated as "clay," the term "clay mineral" being reserved for specific fine-grained layer-lattice aluminosilicates, or groups of related minerals of this type, such as kaolinite. These conventions will be observed here, except that in some instances clay minerals will be called "clay" for convenience, as in expressions like "clay crystals." Oxides (e.g. MnO_2), hydroxides (e.g. $Al(OH)_3$), and oxyhydroxides (FeOOH, MnOOH, and AlOOH) will be lumped together under the rubric "oxides" for brevity. "Heavy metals," which include both transition and nontransition metals, are conventionally defined as metallic elements of specific gravity > 5 g/cm^3 (Martin and Coughtrey 1982; Morris 1992), and "metalloids" are defined as elements possessing both metallic and nonmetallic properties (Morris 1992). Note, however, that heavy metals are more meaningfully distinguished from other kinds of metals, and from each other, by an assortment of properties that have a more direct bearing on the nature of their chemical bonds and reactions (Douglas and McDaniel 1965; see below). Collectively, heavy metals and metalloids will be referred to as "trace elements", and individual elements will be identified by their chemical symbols (Hg, Se, etc.).

5.2 Properties of Clay Minerals and Oxides

Clay minerals and oxides are widespread and abundant in aquatic and terrestrial environments; and, being finely divided, they have large surface areas per unit of mass. Therefore, the biogeochemical and ecological effects of their surface properties and surface reactions are greatly magnified. The surface properties are functions of mineral composition and structure, with important modifications imposed by environmental variables, such as pH.

Clay minerals and oxides differ greatly in composition, structure, and behaviour but are comparable in certain respects. Both types of mineral are fine-grained crystalline or short-range order solids which may be defined as polymeric inorganic complexes comprising arrays of cations, such as Al (III), Si (IV), Fe (III), and Mn (IV), coordinated to shared ligands consisting of O atoms and OH groups, with O and OH units forming the surfaces of the mineral particles (Fig. 5.1). Owing to the presence of multiple surface charges, as well as partial charges due to dipole effects, minerals of both groups bind and exchange cations, anions, and uncharged polar species in aqueous media and may act as catalysts and as centres of nucleation for precipitates. As a result of reversible pH-controlled binding and release of H^+ ions by amphoteric surface OH groups and O atoms, oxides and the *edge* faces of clay crystals have temporary positive and negative charges, and their net surface charge is positive, negative, or zero, depending on the pH of the solution phase in contact with the minerals (Stumm et al. 1976). However, a fundamental distinction between clay minerals and oxides is that clay minerals are layer-lattice aluminosilicates and have 001 (basal cleavage, or planar) faces with permanent negatively charged sites owing to isomorphous cation substitution within the clay structure. The charge density depends on the nature of the clay and is independent of pH; and since 001 faces comprise a much larger total area than edge faces, a clay crystal typically has a net negative charge. Whereas the sorption of ions by 001 faces of clay is limited to loose electrostatic binding of cations, oxides and edge faces of clay bind both cations and anions, often forming strong, partly covalent surface complexes with them. In general, oxides sorb metals much more effectively than clay minerals do. Moreover, in contrast to both Al oxides and clay minerals (except those containing metals such as Fe, Mn, and Cr), Fe and Mn oxides are involved in oxidation–reduction reactions, and

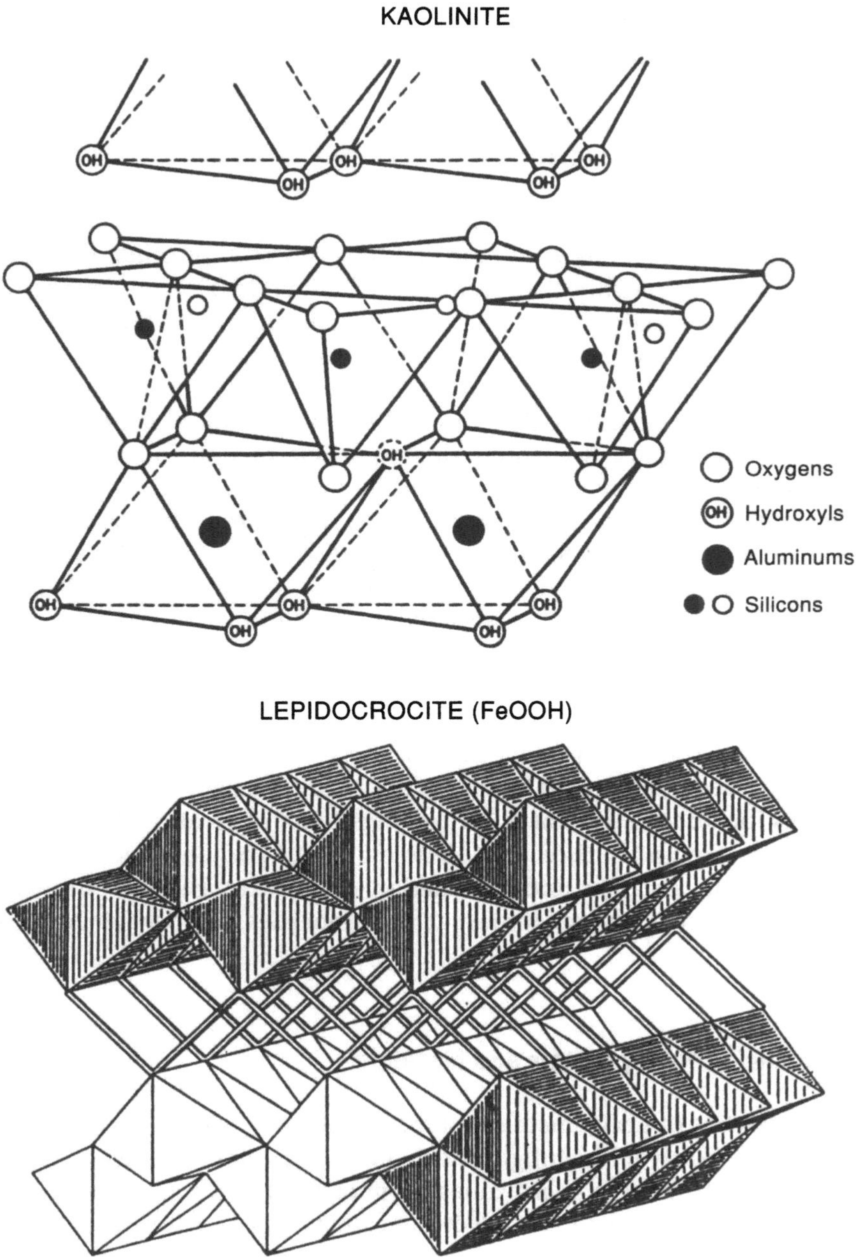

Fig. 5.1. The structures of representative colloidal minerals: the clay mineral kaolinite and the Fe oxyhydroxide lepidocrocite. (From Grim 1968 and Pauling 1960, respectively; republished with the permission of the McGraw-Hill Book Co. and Cornell University Press, respectively)

they precipitate and dissolve in response to increases and decreases, respectively, in the dissolved O_2 content and redox potential (Eh) of the medium (Fe oxide forming at lower O_2 and Eh levels at a given pH than Mn oxide does) (Degens 1965). Unlike clay minerals, oxides readily precipitate or dissolve under earth-surface conditions and are commonly "amorphous" (have short-range order) when freshly precipitated, although they gradually crystallise on ageing; among clay minerals, only allophane is amorphous. Also note that the properties of mineral colloids vary widely even within the same class of minerals. Thus, montmorillonite has higher cation sorption and exchange capacities than kaolinite does because of its finer grain size and larger specific surface, its greater permanent negative charge, and its interlamellar spaces (Grim 1968); and, as we have seen, Al oxide, unlike Fe and Mn oxides, is not affected by redox conditions. Even on a single particle of uniform mineralogy, surface properties vary, as exemplified by the contrast between 001 and edge faces of clay; and there is variation, as well, among sorption sites on a single crystal face. As for grosser characteristics, both oxides and clay minerals are lyophobic colloids which may occur as dispersed individual particles or as aggregates. In an aqueous medium each particle is surrounded by an electrical double layer with a zeta potential that varies with the ion content of the ambient solution, and the particles are subject to flocculation or peptisation, depending on external conditions. Finally, it is necessary to bear in mind that the surfaces of colloidal mineral particles are often altered or masked by interactions with the environment. Surface changes include the deposition of coatings (e.g. layers of oxide or humic matter on clay, and sorbed phosphate on oxides and edge faces of clay) as well as the creation and elimination of surface charges. The result may be a profound change in the surface properties and behaviour of the particles.

5.3 Properties of Heavy Metals and Metalloids

The properties of heavy metals and metalloids vary widely. Owing to differences in electronic configuration, atomic size, and nuclear charge, these elements differ in their values for parameters such as valence, ionic radius, ionisation potentials, electronegativity, polarisability, log K_1 of metal cation hydrolysis, entropy and enthalpy of hydration, standard oxidation–reduction potentials, and (in the case of transition metals) crystal field stabilisation energy (Douglas and McDaniel 1965). Thus, heavy metals vary in the stability and degree of covalent character of bonds that they form with ligands and adsorbents, in their tendency to

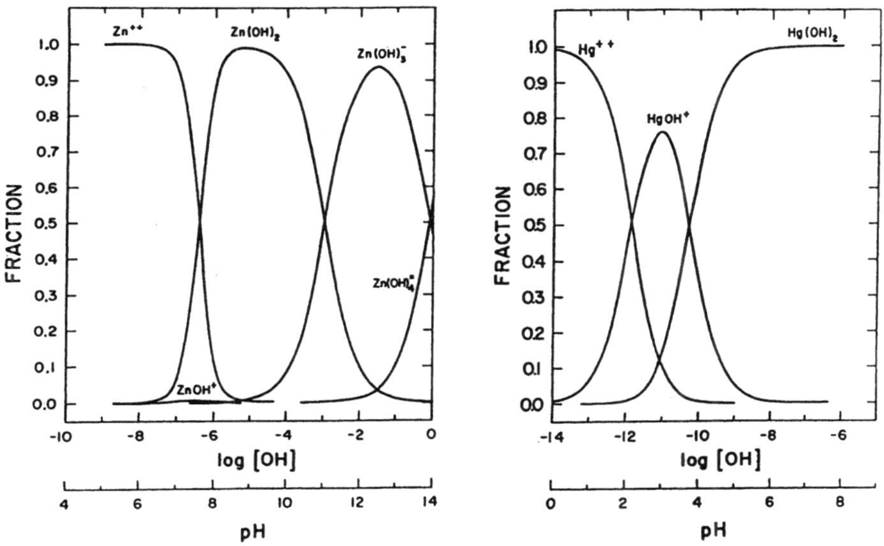

Fig. 5.2. Variations in the proportions of free cations and hydroxylated species of Zn and Hg in aqueous solution as functions of ambient pH. (From Hahne and Kroontje 1973; republished with the permission of the publishers of the Journal of Environmental Quality)

displace one another from binding sites, and in their tendency to be oxidised or reduced under a given set of conditions. An important characteristic of any aqueous heavy metal cation is its strong affinity for OH^- ions, whereby it undergoes stepwise hydroxylations within certain pH ranges which are specific for the metal in question. At sufficiently low pH values (the pH range depending on the individual metal), the dominant species in aqueous solution (assuming that ligands other than H_2O and OH^- are absent) is the "free" aquo cation (e.g. Zn^{2+}, more correctly represented as $Zn(H_2O)_6^{2+}$), but with rising pH it is hydrolysed to an hydroxylated cation (e.g. $HgOH^+$) and then to a simple uncharged complex (e.g. $Zn(OH)_2$) (possibly accompanied by polymeric hydroxylated species), and finally, under alkaline conditions, to anionic hydroxy species (e.g. $Zn(OH)_3^-$ and $Zn(OH)_4^{2-}$) (Fig. 5.2). Among some heavy metals (e.g. Cd and Zn) the free cation is the dominant species throughout the acidic pH range and even under mildly alkaline conditions, but among others (e.g. Hg), the free cation predominates only under extremely acidic conditions (Hahne and Kroontje 1973). Within the pH range of most natural environments (~ 3–10) (Baas Becking et al. 1960), the principal hydrated and hydroxylated forms of most heavy metals in solution are cationic and uncharged

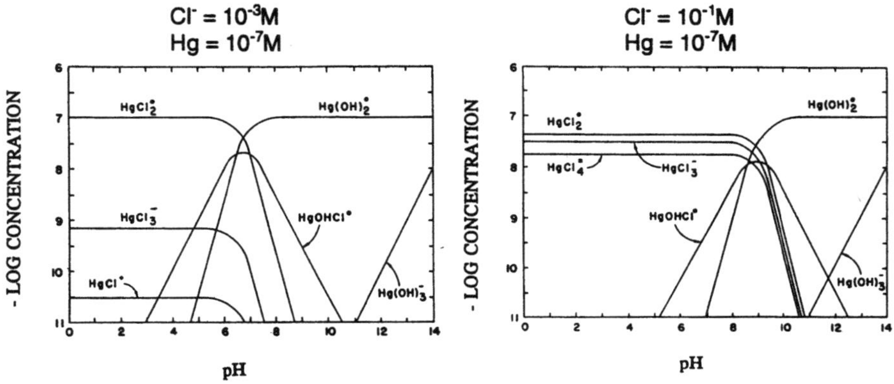

Fig. 5.3. Variations in the concentrations of chloride and hydroxy complexes of Hg in aqueous solution as functions of pH at two different Cl^- levels. (From Leckie and James 1974) (The publishers [Butterworth Heinemann] no longer hold the copyright)

species. However, certain metals (e.g. Cr, Mo, and V), as well as metalloids such as As and Se, occur as complex oxyhydroxy anions (e.g. $H_2AsO_4^-$) at ordinary pH values. Being amphoteric, these anions can either bind or release H^+ ions, depending on the ambient pH. In natural waters, of course, numerous ligands compete with H_2O and OH^- ions for trace elements; for instance, Cl^- ions tend to displace OH^- ions from Hg(II), forming various dissolved complexes such as $HgCl_2$, $HgCl_3^-$, and $HgCl^+$, unless the pH is high enough for OH^- ions to prevail (Fig. 5.3).

Although heavy metals differ greatly amongst themselves, they have broad similarities which set them apart from alkali and alkaline-earth metals. Ionic bonding is characteristic of the latter, but heavy metals have a more or less marked tendency to form covalent bonds, leading to the production of stable complexes and compounds, such as chelate, chloride, hydroxy, oxyhydroxy, and thiol complexes, surface complexes formed at sorption sites on mineral particles, nearly insoluble sulfides, and organometallic compounds. The distinctive behaviour of heavy metals is a consequence of (1) their relatively strong nuclear charge, (2) inefficient shielding of the nuclear charge by filled or partly empty *d* and *f* electron subshells underlying the valence electrons, (3) penetration of these inner subshells by valence electrons, and (4) cation polarisability due largely to *d* and *f* electrons. Because of these factors, the electronegativities and ionisation potentials of heavy metals are relatively high, and energy levels of *d* orbitals in transition metals manifest crystal field splitting in the presence of ligands. Thus, a heavy metal atom has a large "effective nuclear charge", exerting a strong pull on the primary valence electrons and on secondary valence electrons donated by ligands. This favours formation of stable bonds with considerable covalent character. The radically different behaviour of alkali and alkaline earth metals arises from the fact that the electron subshells underlying the valence electrons have "noble gas" configurations, resulting in highly effective shielding of the nucleus, with little or no penetration of inner subshells by valence electrons.

5.4
The Binding, Release, and Transport of Trace Elements by Minerals

Trace elements in natural waters and soil solutions are readily taken up from solution by colloids, the efficiency of the phenomenon depending on the properties and concentrations of the reactants and on environmental factors that affect the forms of the elements and the surface properties of the colloids. This process involves (1) sorption by clay minerals, oxides, and heterogeneous particles (e.g. clay–oxide–humic complexes), (2) coprecipitation with oxides, and (3) complexing by organic (notably humic) matter. Carrying bound trace elements with them, the colloids may be transported by wind or water, settling out under suitable conditions, either as dispersed individual particles or as aggregates formed by flocculation (coagulation) or by gradual growth of composite particles (such as the "stable aggregates" of soil) owing to mutual binding of different substances (e.g. clay and humic matter). Sorbed trace elements may be desorbed under favourable conditions, but desorption is often slow, difficult, and incomplete; and trace elements enclosed in oxide precipitates are probably released only if the precipitates are solubilised. As it is chiefly these phenomena that account for the key significance of colloidal minerals in the biogeochemical cycles of trace elements, we will examine them in greater detail before considering their biological and environmental implications. First we will review the basic processes, which are known largely from reactions between trace elements and individual minerals in simple experimental systems (Sects. 5.4.1, 5.4.2, 5.4.3). Then we will turn to evidence from natural environments and from model systems containing soils or sediments in their native state (Sect. 5.4.4). Although we have only limited ability to extrapolate from simple model systems to complex, less well defined natural ones (Pickering 1979; Anderson 1981; Swift and McLaren 1991), data from laboratory experiments have been used successfully to predict sorption phenomena occurring in nature (Tessier et al. 1996).

5.4.1 Sorption and Desorption

"Sorption" involves adsorption (binding to the surface of a particle) but also implies possible migration of adsorbed species into the interior of the particle, or precipitation of discrete solid phases (e.g. polymerised hydrolysis products) on its surface, or both. Here the term "sorption" will be used simply to denote uptake of dissolved substances without reference to this distinction. "Desorption", the reverse of sorption, has been defined rigorously as release of sorbed species into solution on immersion in a solvent of the same composition (in terms of background solutes) as the solvent from which it was taken up (Swift and McLaren 1991); here, however, the term will be defined loosely as release of sorbed species under any circumstances whatsoever. The sorption and desorption of trace elements by colloidal minerals have been reviewed thoroughly elsewhere (Grim 1968; Jenne 1968; Leckie and James 1974; Lee 1975; Harmsen 1977; Hingston 1981; Kinniburgh and Jackson 1981; Schindler 1981; Bruemmer et al. 1986; Leckie 1986; Schindler and Stumm 1987; Westall 1987; Förstner 1991; McBride 1991b; Schindler and Sposito 1991; Schmitt and Sticher 1991; Schuster 1991; Swift and McLaren 1991), and the merits and limitations of predictive theoretical models of these phenomena have been discussed (Pickering 1979; Anderson 1981; Swift and McLaren 1991). Although there is broad agreement about a number of fundamentals, sorption mechanisms are not fully understood, and different theories have been proposed to explain them (Leckie and James 1974; Pickering 1979; McBride 1991b).

All clay minerals and oxides are able to sorb cationic, anionic, and uncharged metal species and can take part in cation and anion exchange reactions, but they differ greatly in their sorption capacities, their cation and anion exchange capacities, and the binding energies of their sorption sites (Lee 1975; Harmsen 1977; Pickering 1979, 1980a, 1980b; Kinniburgh and Jackson 1981; Schindler 1981; Bruemmer et al. 1986; Brümmer 1986; Leckie 1986; McBride 1991b; Schindler and Sposito 1991; Schmitt and Sticher 1991). Even a single mineralogically homogeneous particle (e.g. a montmorillonite crystal or a speck of amorphous FeOOH) typically has different kinds of sorption sites that span a range of binding energies (Benjamin and Leckie 1981; McBride 1991b). Moreover, binding *capacity* (quantity sorbed per unit mass of sorbent) and binding *energy* (strength of sorption) vary, to a greater or lesser extent, independently of each other, and reported correlations between them range from positive (Shuman 1977) to negative (Siegel and Siegel 1979). Binding capacity presumably depends in part on binding energy but is also a function of other factors, such as specific surface area. Because the sorption of trace elements by mineral colloids involves a number of different binding sites, trace element species, and environmental factors, the binding mechanisms and strength of binding are highly variable (Harmsen 1977; McBride 1991b; Swift and McLaren 1991). Cations and anions may be held rather loosely by exchange sites with negative and positive charges, respectively, or more strongly by charged or uncharged surface ligands. As explained above, oxide surfaces and edge faces of clay have temporary negative and positive charges whose concentrations per unit of surface area vary with the ambient pH, whereas the 001 faces of a clay crystal have fixed numbers of permanent negatively charged sites. Bonds between ions and sorption sites range from ionic to partially covalent, and van der Waals' forces and other dipole effects play a role in the sorption of uncharged species, while H-bonds between the metal's hydration sphere and the mineral surface also contribute to the sorption process (Harmsen 1977). In general, sorption occurs rapidly (although cations take longer to invade the interlamellar spaces of montmorillonite than to cover the outer surfaces (Pickering 1979)); but the binding of cations is often followed by slow diffusion of some of the cations into the interior of the sorbent through minute pores, defects, and dislocations in the crystal structure, a process that may be facilitated by dehydration and may not be readily reversed (Harmsen 1977; Pickering 1979; Bruemmer et al. 1986; Förstner 1991).

5.4.1.1
Cation Sorption and Exchange on 001 Faces of Clay Crystals

Exchange sites for cations comprise points of (1) permanent negative charge on 001 faces of clay minerals and (2) transitory negative charge produced by the reversible, pH-dependent dissociation of H^+ ions from amphoteric surface OH groups of oxides and edge faces of clay minerals (see Sect. 5.4.1.2) (Stumm et al. 1976; McBride 1991b; Swift and McLaren 1991). Cations are sorbed to sites on 001 faces of clay primarily by long-range coulombic attraction (ionic bonding), and they keep their hydration spheres wholly or partially intact so that they are not in direct contact with the clay surface, although H_2O molecules and OH^- groups coordinated to the cations help to anchor the cations to the clay by forming H-bonds with surface O atoms (McBride 1991b; Fig. 5.4). A cation's closeness of approach to the mineral surface, and, therefore, the strength of binding, increase as H_2O is lost from the hydration sphere (McBride 1976). At 001 sites, sorption is more or less readily reversed, and the sorbed cations are subject to cation exchange. The binding energies are relatively low, but the strength of binding is thought to vary depending on

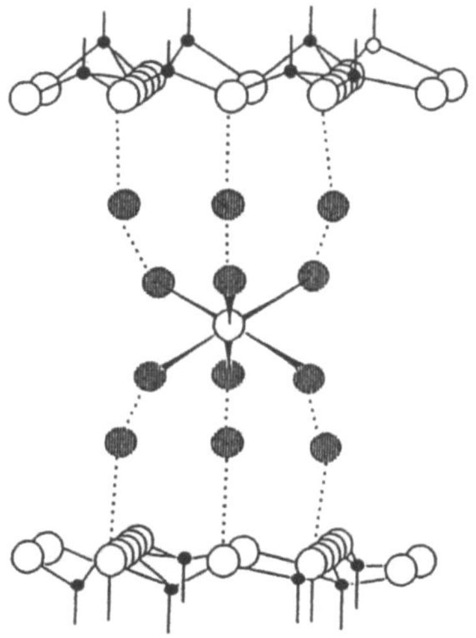

Fig. 5.4. The adsorption of an hydrated divalent metal cation ($M(H_2O)_6^{2+}$) by the interlayer surfaces of montmorillonite. *White ball in centre* Metal cation; *grey balls* water molecules; *upper* and *lower structures* oxygen atoms (*white balls*) coordinated to silicon atoms (*black balls*) of the clay mineral; *dotted lines* H-bonds. (From McBride 1991b; republished with the permission of Kluwer Academic Publishers and M.B. McBride)

whether the isomorphous substitution responsible for the negative charge is in the tetrahedral or octahedral layer of the clay (i.e. nearer to or farther from the surface) (Grim 1968). The tendency of one kind of cation to displace another varies with the nature of the cations and the clay, the concentrations of the cations, and environmental conditions. Thus, cations of high charge, large size, and a low degree of hydration are bound preferentially (except for H^+ ions, which behave like cations of higher valence) (Grim 1968; McBride 1976, 1980, 1991b). In addition, relative affinities of clay minerals for metals depend on the nature of the mineral and the experimental conditions (Pickering 1980a, b). However, free divalent cations of heavy metals have the same affinity for exchange sites of 001 faces of clay as alkaline-earth cations of similar ionic radius (McBride 1991b). Except at pH values so low that dissolved heavy metals are mostly in the form of free cations, sorption of heavy metal cations to 001 exchange sites of clay is probably of scant significance in nature, because these cations tend to be displaced by the much more abundant "major cations" (Ca^{2+}, Mg^{2+}, Na^+, etc.) owing to the Law of Mass Action (Abd-Elfattah and Wada 1981; Swift and McLaren 1991), although differences in metal ion properties cause deviations from the predicted results (McBride

1980). But as the pH drops, competition between H^+ ions and metal cations for exchange sites becomes increasingly important (Pickering 1979, 1980b). Thus, heavy metals are probably not extensively sorbed by 001 faces of clay at any pH simply because their cations are always vastly outnumbered by other cations with which they compete.

5.4.1.2
"Specific Sorption" and Exchange of Cations, Anions, and Uncharged Species by Oxides and Edge Faces of Clay

Unlike cations sorbed to basal surfaces of clay, cations sorbed by oxides and by edge faces of clay are bound directly to surface OH groups and O atoms (including O^- formed by dissociation of H^+ from OH), which function as ligands, displacing monodentate ligands such as H_2O from dissolved metal cations by ligand exchange and forming bridges linking the cations to structural cations in the mineral (Stumm et al. 1976; Harmsen 1977; Pickering 1979; Leckie 1986; McBride 1991b; Schuster 1991; Swift and McLaren 1991; Zachara et al. 1991; Hsia et al. 1992; Stadler and Schindler 1993a; Figs. 5.5, 5.6A). The bonds are intermediate between ionic and covalent (McBride 1991b), and are stronger than the coulombic forces that bind cations to 001 exchange sites of clay (Stumm et al. 1976); moreover, the geometry of the OH groups of oxides favours the formation of coordination polyhedra with metals at the mineral surfaces (Manceau et al. 1992). Sorption, or chemisorption (Harmsen 1977), of metal cations by surface ligands of oxides and edge faces of clay has been aptly called "surface complexation" but is usually known as "specific sorption"; both terms will be used interchangeably here. An oxide or clay mineral is essentially a gigantic polymeric coordination complex, and, by sharing surface ligands with it, heavy metal cations extend its structure (Fig. 5.6A). Some involvement of 001 faces as well as edge faces of clay has been postulated (Swift and McLaren 1991), but the evidence for it is indirect and unconvincing. The importance of surface complexing in heavy metal geochemistry is demonstrated by the fact that Fe, Mn, and Al oxides have higher sorption capacities for Zn than montmorillonite does owing to their higher concentrations of surface ligands, even though montmorillonite, whose 001 faces account for most of its surface area, has a higher cation exchange capacity (Brümmer et al. 1983; Table 5.1). Also note that surface complexing alters the distribution of charges on the surface of the colloid; thus, heavy metal cations sorbed by kaolinite may change the net charge on the mineral from negative to positive (Hunter and James 1992). A change in surface charge could lead to effects such as flocculation or binding of the clay by negatively charged ligands

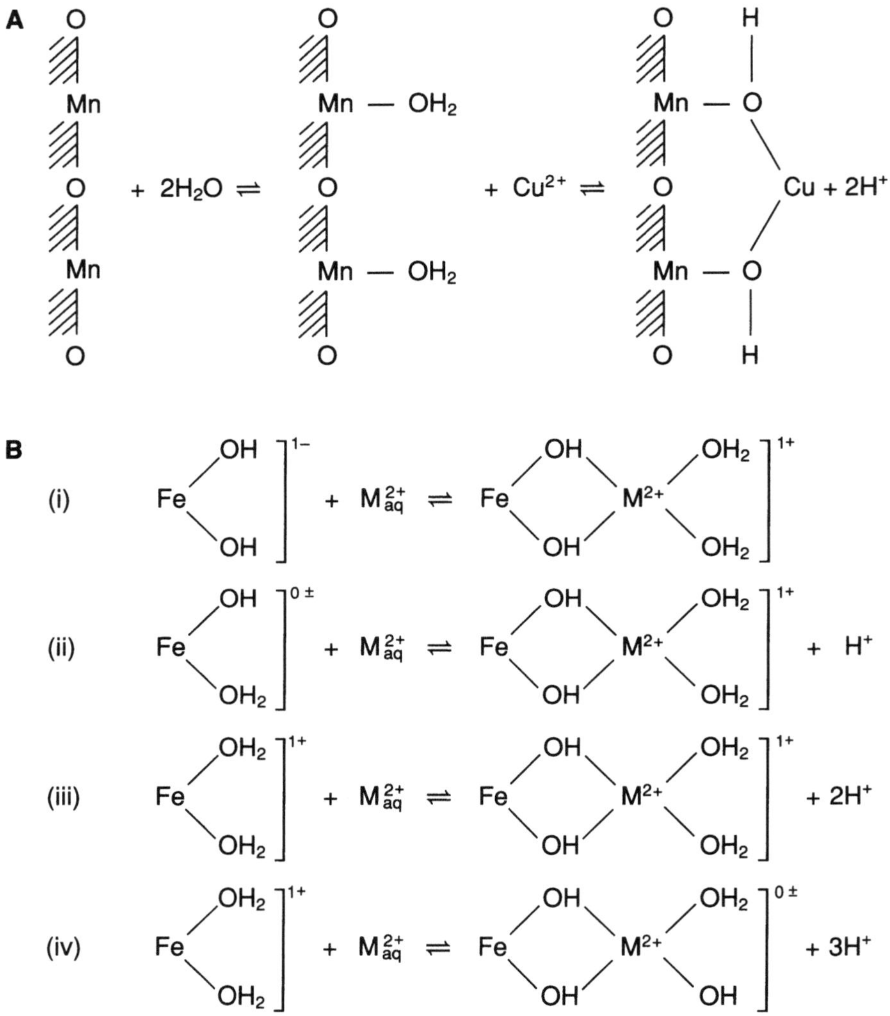

Fig. 5.5 A, B. The specific sorption of divalent heavy metals by **A** Mn oxide and **B** Fe oxide. (From Pickering 1979 and Swift and McLaren 1991 [after Kinniburgh 1983], respectively; republished with the permission of John Wiley & Sons, Inc. and of Kluwer Academic Publishers and R. S. Swift, respectively)

(mainly $-COO^-$ and $-O^-$) of humic matter, with sorbed cations acting as a "cement".

Surface complexation may not be easily or completely reversible (Fig. 5.7), but this probably reflects a high activation energy and low rate of desorption rather than true irreversibility (McBride 1991b). Pickering (1979) reported complete recovery of sorbed Cu from clay by extraction

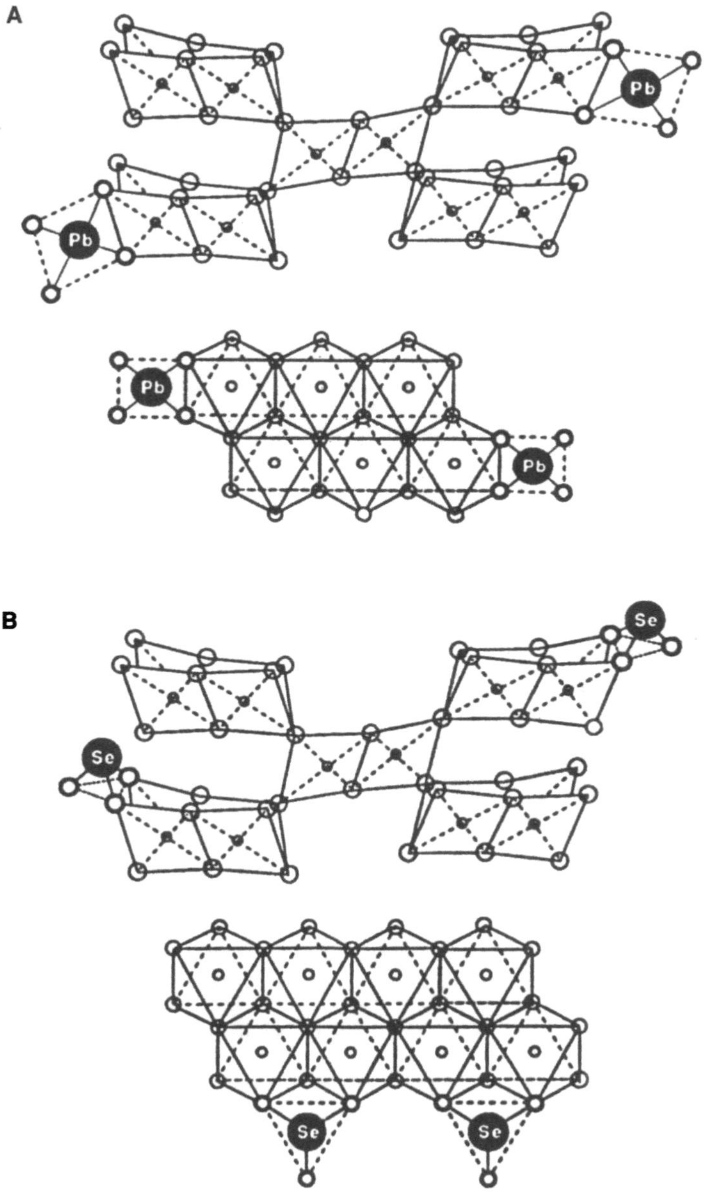

Fig. 5.6 A, B. The specific sorption of **A** Pb^{2+} and **B** SeO_3^{2-} by Fe oxyhydroxide. (From Manceau et al. 1992; republished with the permission of Elsevier Science-NL and A. Manceau)

Table 5.1. Values for the cation exchange capacity and Zn sorption capacity of different naturally occurring sorbents in $CaCO_3$-buffered experimental systems. (Data from Brümmer et al. 1983)

Sorbent	Cation exchange capacity at pH 7.6 (μmol/g)[a]	Zn sorption capacity (μmol/g)[b]
$CaCO_3$	-	0.44
Bentonite	450	44
Humic acid	1700	842
Amorphous Fe oxide	160	1190
Amorphous Al oxide	50	1310
δ-MnO_2 (birnessite)	230	1540

[a] Based on ability of Mg^{2+} to displace Ba^{2+} from Ba^{2+}-saturated sorbent.
[b] Calculated from equations for Langmuir isotherms.

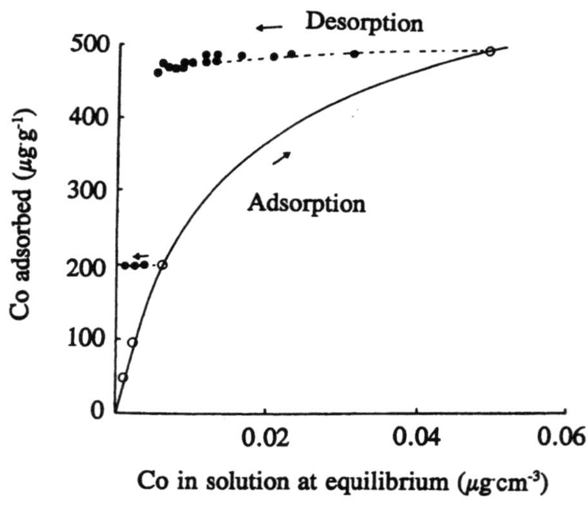

Fig. 5.7. Curves representing adsorption of Co by a naturally occurring Fe-Mn oxide followed by desorption. (From McLaren et al. 1986; republished with the permission of Blackwell Science Ltd. and R. G. McLaren)

with EDTA, proving that specific sorption can be reversed. Surface complexation may or may not involve cation exchange and bonding by negatively charged sites, and so cation sorption is not necessarily balanced by simultaneous desorption of an equivalent number of previously sorbed metal cations. Nevertheless, different heavy metals undoubtedly displace one another from surface ligands, and in any case the binding of metal cations is accompanied by displacement of H^+ ions from OH groups of the sorbent (Leckie and James 1974; Stumm et al. 1976; Pickering 1980a, b;

Leckie 1986; McBride 1991b; Swift and McLaren 1991; Stadler and Schindler 1993a; Fig. 5.5), although some of the H^+ ions may arise by dissociation of H_2O in the hydration spheres of sorbed cations, and specific sorption sites differ in their relative affinities for metal cations and H^+ (Pickering 1979). As these reactions involve cation exchange, terminology that makes a distinction between "cation exchange" and "specific sorption", is inappropriate. Evidently some surface complexation reactions involve cation exchange whereas others do not (Stadler and Schindler 1993a). In other words, both negatively charged and uncharged oxygenic ligands take part in the surface complexing of heavy metal cations. Besides, the term "specific sorption" is misleading, because cation exchange on 001 faces of clay also involves a degree of selectivity based in part on cation properties. Yet cation exchange is commonly described as non-selective (Harmsen 1977; Bruemmer et al. 1986; McBride 1991b; Swift and McLaren 1991). It would be more correct to say that cation exchange on 001 faces of clay is much less selective than specific sorption.

In any case, it must be kept clearly in mind that there is a fundamental difference between (1) the weak but long-range ionic bonding of hydrated cations to negatively charged 001 faces of clay and (2) the much stronger, partly covalent direct bonding of cations by both negatively charged and uncharged surface ligands of oxides and edge faces of clay. These are the two basic mechanisms of cation sorption. It is true, of course, that dissolved heavy metal cations are attracted to negatively charged sites on oxides and edge faces of clay by coulombic forces, as in the sorption of cations by 001 faces of clay; but once the cations reach these sites they form surface complexes with partly covalent bonds, which cannot be formed at sites on 001 faces of clay. Therein lies the essential contrast between the two processes.

Surface ligands (unlike cation exchange sites of 001 faces of clay) have a strong preferential affinity for heavy metals with respect to alkali and alkaline-earth metals (Harmsen 1977; Madrid et al. 1991), besides forming more stable bonds with some heavy metals than with others owing to differences in specific metal properties, such as electronegativity, polarisability, first hydrolysis constant, ionic radius, and entropy and enthalpy of hydration. These properties are related to the metal's tendency to form covalent bonds with ligands, to the hydrated cation's distance of closest approach to the mineral surface, and to steric factors which make it more feasible for some metals than others to fit into openings in the crystal structures of minerals (Harmsen 1977; Rickard and Nriagu 1978; Pickering 1980b; Bruemmer et al. 1986; Brümmer 1986; Schmitt and Sticher 1991). Moreover, the ability of colloids to sorb metal ions is affected by the the nature and abundance of other metal ions already bound to the sorption sites (Pickering 1980a). Above all, selective sorption correlates with metal hydrolysis, which is a function of pH (see Sect. 5.4.1.4); thus, hydroxylated metal cations (e.g. $CuOH^+$) appear to be sorbed preferentially with respect to free cations (e.g. Cu^{2+}), possibly because they are less strongly hydrated (Leckie and James 1974; McBride 1991b; Schuster 1991). However, it is not known whether the hydroxy cation forms in solution and is then sorbed or whether the hydrated free cation is sorbed and then releases an H^+ ion from its hydration sphere (Leckie and James 1974; Pickering 1979); a third possibility is that the OH group of an hydroxy cation can strip an H^+ ion from an OH group of the mineral, creating a negatively charged site to which the metal (now converted back to a free cation) is then bound (Pickering 1980a, b). Another possible, though unproved, mechanism of sorption is precipitation of a metal hydroxide phase on the mineral surface under circumneutral or alkaline conditions, but this is possible only if the solution phase is saturated or supersaturated with the metal (Leckie and James 1974; Pickering 1979; Bruemmer et al. 1986; Brümmer 1986; Förstner 1991; Swift and McLaren 1991). In short, specific sorption is not fully understood. It may involve several different mechanisms, and the sorbed metal species may range from simple cations to complex polynuclear hydroxides of variable composition formed by hydrolysis and condensation in solution or by nucleation on the mineral surface (McBride 1991). But at the low heavy metal concentrations found in most natural waters, the dominant process is probably adsorption, not precipitation (Bruemmer et al. 1986; Brümmer 1986).

Oxides and edge faces of clay bind and exchange anions as well as cations. Anion exchange sites are points of temporary positive charge formed by reversible pH-controlled association of H^+ ions with amphoteric OH and O (Stumm et al., 1976; Swift and McLaren 1991), or by displacement of OH^- (Grim 1968; Hingston et al. 1968), on the mineral surfaces. In the specific sorption of heavy metal and metalloid anions (as in the specific sorption of heavy metal cations), ligand exchange reactions probably occur, and oxygenic ligands (OH groups and O atoms) act as bridges linking structural cations (e.g. Fe^{3+} and Al^{3+}) of oxides and edges of clay crystals to the metal or metalloid of the anion, forming a binary coordination complex at the mineral surface (Hingston et al. 1968; Swift and McLaren 1991) – in effect, an extension of the mineral (Grim 1968) (Fig. 5.6B). Anions may displace OH^- groups from structural cations of the minerals (Grim 1968; Hingston et al. 1968), but displacement of OH^- from the anion by OH and O coordinated to the structural cations may be equally possible. Dissolved anionic species of heavy metals and metalloids are attracted to positively charged sites by coulombic forces but are subsequently fixed to these sites by partly covalent coordinate bonds and the sharing of oxygenic ligands with structural cations of the mineral, as described above.

As with cations and anions, uncharged heavy metal species such as $Hg(OH)_2$ are probably subject to surface complexation (Schuster 1991). The principles of coordination chemistry apply equally, regardless of whether the dissolved metal species are positively or negatively charged or uncharged.

5.4.1.3
Physical Sorption of Uncharged Species

Uncharged heavy metal species, including polynuclear hydroxides formed by hydrolysis as well as simple species such as $Hg(OH)_2$, may be sorbed to uncharged sites of mineral colloids by H-bonding and by van der Waals' (London) bonds (Harmsen 1977) and other dipole interactions. Compared with surface complexation, "physical sorption" by mechanisms such as this is rather weak. As discussed above, precipitation of polymeric metal-OH complexes on colloid surfaces is possible (though unlikely in natural environments) under circumneutral or alkaline conditions (Leckie and James 1974; McBride 1991b).

5.4.1.4
Effects of pH

As we have seen, H^+ ions compete with free metal cations for cation exchange sites at low ambient pH. Where heavy metals are concerned, however, pH is especially important because it controls metal cation hydrolysis and specific sorption. Specific sorption mainly involves heavy metals whose dissolved cations are readily hydrolysed under the conditions prevailing in most natural waters and soil solutions; these metals include many biologically important elements, such as Cu, Zn, Cd, Hg, and Pb. Sorption is controlled by pH in a manner which suggests that hydroxy cations (e.g. $HgOH^+$) and other hydroxylated metal species formed by hydrolysis (a pH-dependent process) are usually bound much more strongly than the free aquo cations of the metals (e.g. Hg^{2+}) (Leckie and James 1974; Forbes et al. 1976; Pickering 1979; Barrow et al. 1981; Bruemmer et al. 1986; McBride 1991b; Schuster 1991; Swift and McLaren 1991), although Barrow et al. (1981) reported preferential sorption of Pb^{2+}, not $PbOH^+$, by goethite. Under sufficiently acidic conditions (the pH range depending on the particular metal), the metals are in the form of free cations, and the percent sorbed by a given quantity of mineral colloid is minimal [although some specific sorption of metal cations by

FeOOH occurs even at pH values so low that the extent of cation hydrolysis is ⩽0.1% (Kinniburgh et al. 1976) and the mineral surface is positively charged (Thornton 1979)]. Probably the weakened tendency of the minerals to sorb metals at such low pH values can be explained in large part by competition between H^+ ions and metal cations for binding sites, and by an increase in the number of cation-repelling positively charged sites, as well as a decrease in the number of negatively charged sites, with decreasing pH owing to uptake of H^+ ions; it is widely believed, however, that selective surface complexing of the hydroxylated cations formed at higher pH values is the main reason, although this interpretation has been challenged (Stumm et al. 1976; Pickering 1979, 1980b). A possible explanation for the postulated preferential sorption of the hydroxylated cation is that its hydration sphere is less stable than that of the free cation and therefore does not hinder surface complexation as much (Leckie and James 1974).

With increasing pH within a narrow range of pH values that is characteristic of the particular metal, the concentration of hydroxy metal cations rises sharply at the expense of free cations, and the percentage of the metal that is bound to the sorbent increases correspondingly, rising abruptly from its minimum to its maximum (Leckie and James 1974; Payne and Pickering 1975; Davis and Leckie 1978; Rickard and Nriagu 1978; Pickering 1979; Leckie et al. 1980; Benjamin and Leckie 1981; Bruemmer et al. 1986; Calder 1988; Madrid et al. 1991; McBride 1991b; Schmitt and Sticher 1991; Swift and McLaren 1991; Manceau et al. 1992; Figs. 5.8, 5.9, and 5.10), while desorption of previously sorbed metals is inhibited (Duarte et al. 1991). With a further increase in pH, uncharged hydroxy species [e.g. $Hg(OH)_2$] become quantitatively significant, and the percentage of the metal bound to the sorbent levels off or peaks; and, with attainment of more or less strongly alkaline conditions, the percentage may decline owing to the formation of an anionic metal species [e.g. $Hg(OH)_3^-$] (Lockwood and Chen 1973; Leckie and James 1974; Inoue and Munemori 1979; Thanabalasingam and Pickering 1990) accompanied by the creation of anion-repelling negative sites on the mineral surface by removal of H^+ ions from OH groups. Occasionally, exceptions to these tendencies have been reported. Farrah and Pickering (1978) found that sorption of Hg(II) by kaolinite and illite was virtually independent of pH within the pH range 3–12, whereas the proportion of the Hg(II) taken up by montmorillonite decreased with rising pH. These aberrant results were ascribed to the stability of $Hg(OH)_2$ and a decrease in the abundance of cationic Hg species with increasing pH. However, the drastic deviation of the montmorillonite data from the kaolinite and illite data suggests that the surface chemistry of the clay was of decisive impor-

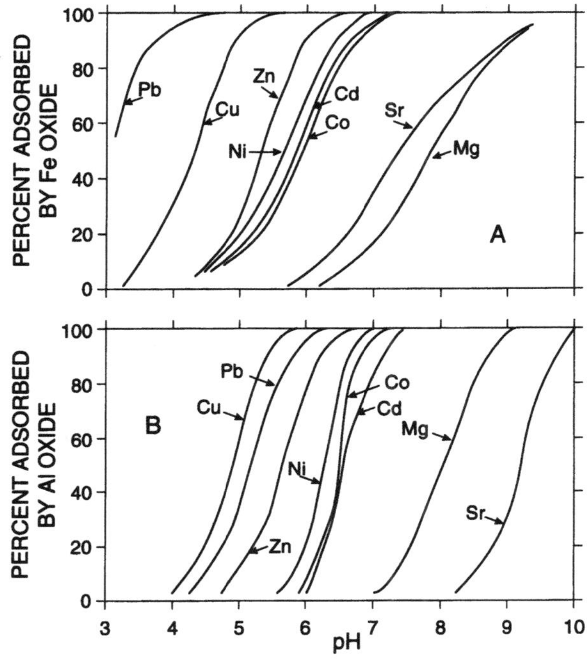

Fig. 5.8 A, B. Effects of pH on the adsorption of divalent metal cations by **A** Fe oxide and **B** Al oxide gels. (From Kinniburgh and Jackson 1981; republished with the permission of D. G. Kinniburgh) (The publishers [Butterworth Heinemann] no longer hold the copyright)

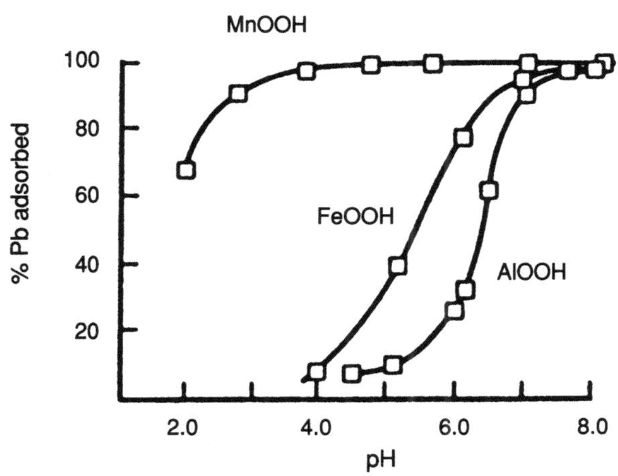

Fig. 5.9. Effects of pH on the adsorption of Pb by Mn, Fe, and Al oxyhydroxides. (From Gadde and Laitinen 1974; republished with the permission of the American Chemical Society)

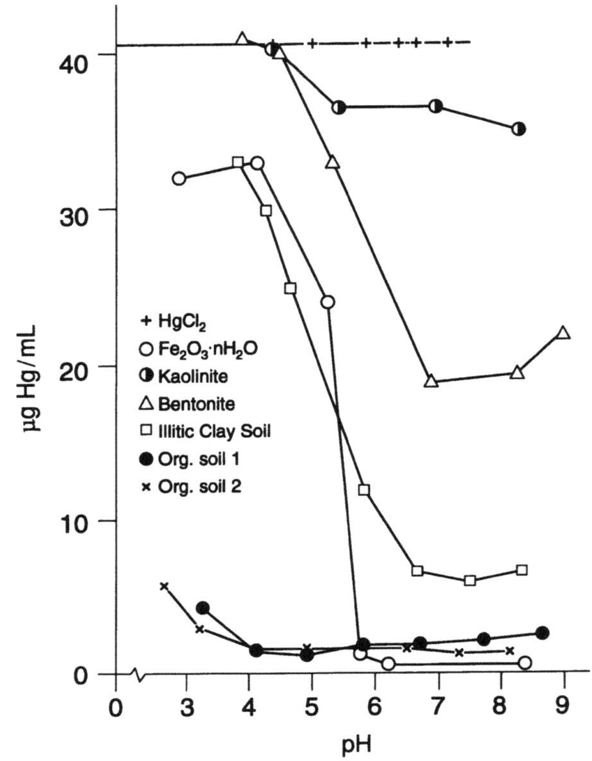

Fig. 5.10. Effects of pH on the sorption of aqueous Hg(II) (in the form of $HgCl_2$) by different soils and soil constituents as represented by Hg(II) concentration remaining in solution plotted against pH of solution. *Org.* organic. (From Andersson 1979; republished with the permission of J. O. Nriagu (editor) and A. Andersson) (The publishers [Elsevier/North-Holland Biomedical Press] no longer hold the copyright)

tance. As the clay specimens had been "finely ground" and subjected to other treatments before being used in the experiments, the meaning of the results is open to question.

From the fact that metals are hydrolysed within different pH ranges, it is apparent that in a mixed solution of metals the influence of pH may be responsible for preferential binding of some metals with respect to others, thereby altering the proportions of the metals remaining in solution (Kinniburgh et al. 1976; Kinniburgh and Jackson 1981). But this is an oversimplification, as other factors besides the pH range of cation hydrolysis are involved. Thus, the relative affinities of metal cations for mineral surfaces at different pH values also depend on the nature of the mineral (Kinniburgh et al. 1976). Pickering (1979, 1980 a,b) has pointed out that the observed pH effect is not due solely to the formation of hydrolysis products of metals and that other possible mechanisms include removal of H^+ ions from potential sorption sites and a decrease in the number of H^+ ions that can compete with free metal cations (Schindler and Stumm 1987). Stumm et al. (1976) went so far as to deny the need to invoke metal hydrolysis at all, suggesting as an alternative the binding of free metal cations by negatively charged sorption sites. The situation is made all the more confusing by evidence that the microenvironments at clay surfaces inhibit metal hydrolysis at high pH (because they are more acidic than the bulk solution) but promote hydrolysis at low pH (McBride 1991b); moreover, cations at exchange sites in the interlamellar spaces of montmorillonite are hindered from reacting with OH^- ions to form hydrolysis products (McBride 1991b). Perhaps the observed overall effect of pH on sorption is a complex function of all of these factors, and others besides, operating simultaneously to varying degrees. Whatever the mechanisms involved, it is possible to state, as a general rule, that the sorption capacity of a colloidal mineral for readily hydrolysable metal cations increases sharply when an acidic solution of the metal is neutralised.

The sorption of anions, too, is a function of pH. Thus, with increasing pH within a certain range, sorption decreases or else increases to a maximum and then decreases (Hingston et al. 1968; Swift and McLaren 1991). In general, sorption of anions is most efficient under highly acidic conditions, which maximise the number of positively charged sorption sites and minimise the number of negatively charged ones, whereas it is least efficient under highly alkaline conditions, which tend to suppress the formation of positively charged sites while fostering the formation of negatively charged ones (Calder 1988).

5.4.1.5
Desorption

It must be kept clearly in mind that desorption is just as important as sorption from an ecological and biogeochemical perspective. Metals differ widely in their ease of desorption as well as their relative affinities for sorption sites (Flemming et al. 1990). Accordingly, some heavy metals sorbed to clay minerals and oxides are much more strongly retained than others, and sorption may not be completely (or at least readily) reversible (Fig. 5.7), although it is not yet possible to generalise about trends among different colloids and metals (Swift and McLaren 1991). Partial or complete desorption can be achieved by immersing a metal-sorbent complex in a metal-poor solution, and by lowering or raising the pH (depending on whether the sorbed ions are cations or anions), by lowering the Eh to solubilise the sorbent (in the case of Fe and Mn oxides), and by the addition of competing ions or complexing agents to the ambient solution.

5.4.1.6 *Effects of Complexing Agents*

Organic and inorganic complexing agents may either hinder or promote the sorption of metal cations, depending, respectively, on whether the metal complexes which they form remain in solution or are themselves sorbed by the mineral, and on whether sorbed complexes are bound more strongly or less so than the free cations (Davis and Leckie 1978; Farrah and Pickering 1978; Leckie and Davis 1979; Pickering 1979, 1980 a, b; Kinniburgh and Jackson 1981; McBride 1991b; Puls et al. 1991; Stockmeyer and Kruse 1991; Swift and McLaren 1991; Sakurai and Huang 1995). But some complexing agents have no effect (or no *net* effect) on metal sorption, at least under certain conditions (e.g. within a given pH range) (Pickering 1980 a, b). The effects of complexing are complicated, as they depend on specific properties of the complexing agents, metals, and sorbents, and on various environmental factors, such as the pH and ionic strength of the aqueous phase. Therefore, each case has to be examined individually (Pickering 1980 a).

Anionic complexing agents, including organic acids and Cl^- ions, often prevent or reverse the association of metal cations with negatively charged sorbents such as clay minerals by forming stable dissolved or dispersed negatively charged complexes with the cations (Feick et al. 1972; Reimers and Krenkel 1974; Lee 1975; García-Miragaya and Page 1976; Newton et al. 1976; Stumm et al. 1976; Pickering 1979, 1980 a,b; Leckie et al. 1980; Kinniburgh and Jackson 1981; Jackson et al. 1982; Wang et al. 1985,

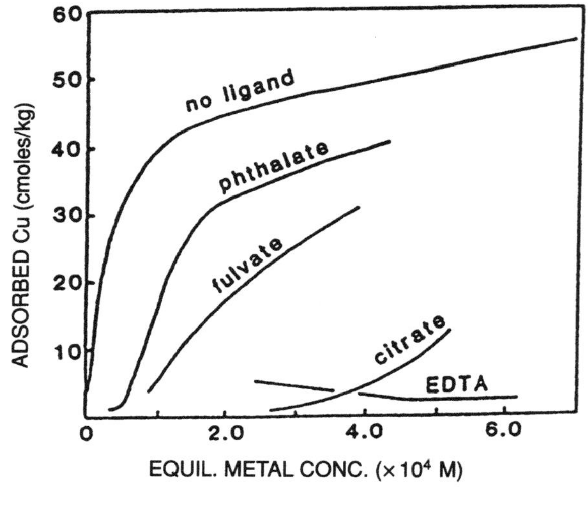

Fig. 5.11. Effects of various organic complexing agents on the adsorption of Cu by montmorillonite. (From McBride 1991 b; republished with the permission of Kluwer Academic Publishers and M. B. McBride)

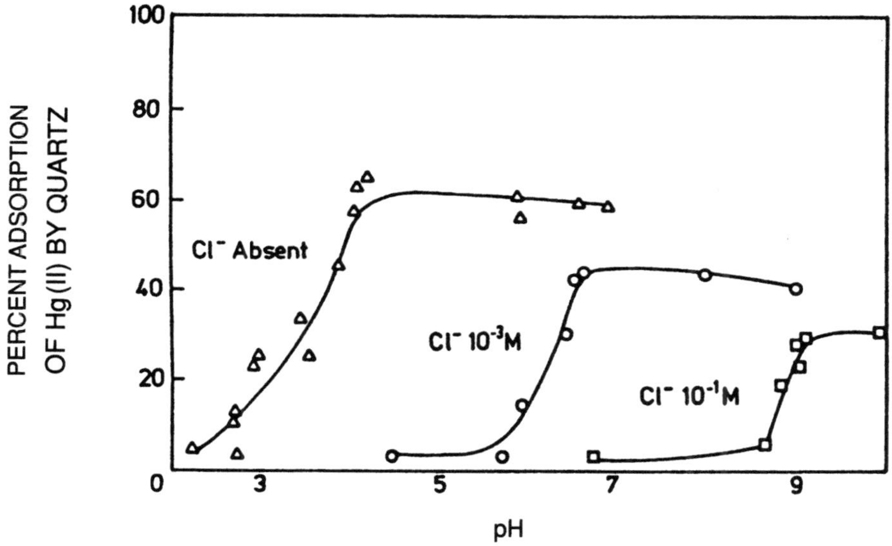

Fig. 5.12. Effects of pH on the adsorption of Hg(II) by quartz in the presence and absence of different concentrations of dissolved Cl^-. (From Schuster 1991, after Leckie and James 1974; republished with the permission of Kluwer Academic Publishers)

1988, 1991; McBride 1991b; Schuster 1991; Erel and Morgan 1992; Roy et al. 1993) (Figs. 5.11, 5.12). Their effect is two-fold: they compete with the minerals for the cations, and the complexes remain separated from the minerals by electrostatic repulsion. Alternatively, the complexes themselves may be sorbed, although their affinity for the mineral surfaces is not necessarily the same as that of the free metal cations (Pickering 1980b). Thus, complexing agents may increase the proportion of the metal taken up from solution if the complex is subject to sorption (Leckie and James 1974), as with positively charged complexes sorbed by clay (Kinniburgh and Jackson 1981; McBride 1991b). Some simple organic acids that function as chelating agents (e.g. citrate and oxalate) may enhance metal sorption (Payne and Pickering 1975; Sakurai and Huang 1995); but deprotonated organic acids (including citrate and oxalate) hinder the sorption of anionic trace element species, such as selenite, by competing with them for positively charged binding sites (Dynes and Huang 1995). Neutral complexes are sorbed by van der Waals' bonding and other weak forces (dipole interactions and H-bonds) (Harmsen 1977) or else are not sorbed (McBride 1991b).

Interactions of humic matter with metals and colloidal minerals are complex, and the effects of humic matter on the sorption of metals by

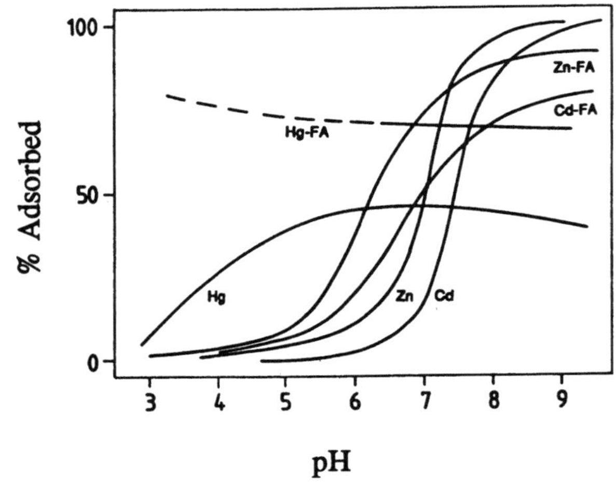

Fig. 5.13. Effects of pH on the adsorption of Zn, Cd, and Hg by Al_2O_3 in the presence and absence of fulvic acid (*FA*). (From Xu and Allard 1991; republished with the permission of Kluwer Academic Publishers)

mineral particles may depend on environmental parameters such as pH (Fig. 5.13). Rich in weakly acidic –COOH and phenolic –OH groups, humic matter usually bears a net negative charge and comprises some of the most effective, abundant, and widespread natural complexing agents. Fulvic acids may interfere with the sorption of metals by clay (McBride 1991b), probably owing to their high solubility in water (resulting in solubilisation of the metals that they bind) and their negative charge (resulting in electrostatic repulsion by clay). Nevertheless, both fulvic and humic acids, along with their chelated metals, are commonly bound to clay minerals and oxides (Pickering 1979), although the higher molecular weight humic fractions may be sorbed preferentially in at least some cases (Marsi 1992). Humic matter is bonded to minerals by bridging mechanisms whereby ligands of the humic matter are coordinated to polyvalent metal cations at the mineral surface, and by H-bonding and van der Waals' forces (and other dipole effects?) (Greenland 1971; Rashid et al. 1972; Schnitzer and Khan 1972; Tipping 1981; Aiken et al. 1985; Hayes and Himes 1986). Similarly, phosphate may enhance the sorption of metals by forming bridges between the mineral surface and the metal (Payne and Pickering 1975; Pickering 1979). In some circumstances, however, phosphate may hinder metal sorption (Du 1992). Of course, the interactions between complexing agents, mineral colloids, and metals are modified by environmental variables, such as pH, which affect surface charges of colloids, protonation and deprotonation of acidic ligands, metal speciation, the strength and mechanisms of binding by ligands and sorption sites, and so on. Besides exerting the effects discussed in Section 5.4.1.4 (above), pH plays an important part in the behaviour of complexing agents. Thus, because humic matter complexes metals mainly by means of acidic ligands whose H^+ ions are split off when the metals are bound (Schnitzer and Khan 1972), the stability of the complexes is higher under alkaline than under acidic conditions; but in the acidic pH range humic complexes are more efficiently sorbed by clay and oxides, probably because the negative charges of the humic matter are largely neutralised by protonation of acidic functional groups (Schnitzer and Khan 1972; Tessier et al. 1996). Accordingly, variations in environmental factors and specific properties of the reactants may well account for apparent paradoxes, as in the opposing effects of fulvic acids (Stadler and Schindler 1993b). Additional complications exist because complexing agents, like mineral colloids, bind metals preferentially. For instance, Cl^- desorbs Hg(II) more effectively than Cu(II) from minerals because it forms more stable dissolved anionic complexes with Hg(II) (Schuster 1991), and the relative affinities of heavy metals for humic matter may depend on any one of several metal properties, including electronegativity (Jackson et al. 1980a; Pickering 1980a, b) and ionisation potentials (Swift and McLaren 1991), or may conform to the Irving-Williams series of complex stabilities (Douglas and McDaniel 1965; Schnitzer and Khan 1972). In any event, the relative affinities of humic matter and other complexing agents for metals are variable and are not necessarily the same as those of mineral colloids, which in turn differ among themselves in that regard. Therefore, the metal preferences of composite colloids such as clay-humic complexes may be difficult to predict and interpret.

5.4.1.7
The Relative Effectiveness of Different Sorbents

Jenne (1968) and Lee (1975) maintained that Fe and Mn oxides are the most important binding agents for heavy metals in soil and aquatic environments, and subsequent work has confirmed and expanded their conclusions. Thus, a body of evidence accumulated by different workers has established that Fe, Mn, and Al oxides, and usually organic matter too, have much higher sorption capacities for heavy metals and metalloids than clay minerals do (Andersson 1979; Huang and Liaw 1979; Förstner and Patchineelam 1980; Pickering 1980a; Abd-Elfattah and Wada 1981; Brümmer et al. 1983; Bruemmer et al. 1986; Kester et al. 1986; McLaren et al. 1986; Jackson 1989; Nielsen 1990; Förstner 1991; Schmitt and Sticher 1991; Table 5.1; Figs. 5.10 and 5.14). This important generalisation has been strikingly illustrated by comparing the Zn sorption capacities of amorphous Fe and Al oxides, δ-MnO_2, bentonite (i.e. montmorillonite), and humic acid in a $CaCO_3$-buffered medium (Brümmer et al. 1983; Bruem-

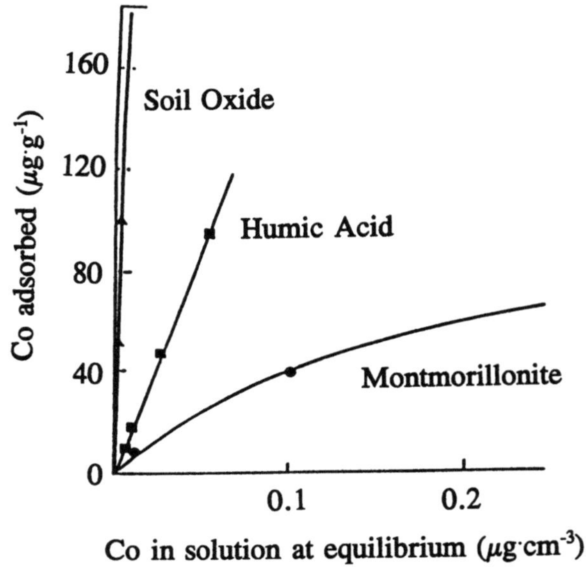

Fig. 5.14. Curves representing adsorption of Co by naturally occurring Fe-Mn oxide ("soil oxide"), humic acid, and montmorillonite. (From McLaren et al. 1986; republished with the permission of Blackwell Science Ltd. and R. G. McLaren)

mer et al. 1986; Brümmer 1986; Förstner 1991). The results show that the cation exchange capacity varied in the order humic acid ⩾ bentonite > oxides, whereas the sorption capacity for Zn varied in the order oxides > humic acid ⩾ bentonite (Table 5.1). Although bentonite had a higher cation exchange capacity (at pH 7.6) than any of the oxides, its sorption capacity for Zn was two orders of magnitude less. There are indications, too, that sorption capacity tends to correlate positively with strength of binding (John 1972; Thornton 1979; Wang et al. 1991; Atanassova 1994; Xing et al. 1995), although this is not necessarily true (Siegel and Siegel 1979). In short, freshly precipitated oxides, and to a lesser extent humic matter, are much more effective scavengers of heavy metals than even the most efficient sorbents among clay minerals, indicating that specific sorption and other complexation processes are the dominant binding mechanisms. However, clay particles play important roles as carriers of associated oxides and humic matter (Kester et al. 1986).

Although oxides as a whole are the most effective sorbents of heavy metals, they differ appreciably among themselves in their sorption capacities (Förstner and Patchineelam 1980; Brümmer et al. 1983; Bruemmer et al. 1986; Brümmer 1986; Wang et al. 1988; Jackson 1989; Thanabalasingam and Pickering 1990; Förstner 1991; Manceau et al. 1992; Fig. 5.9), and in their tendency to release sorbed metals (Rogers et al. 1984). They also show marked preferences for particular metals, as revealed by comparisons between Fe and Mn oxides (Burns and Fuerstenau 1966; Calvert and

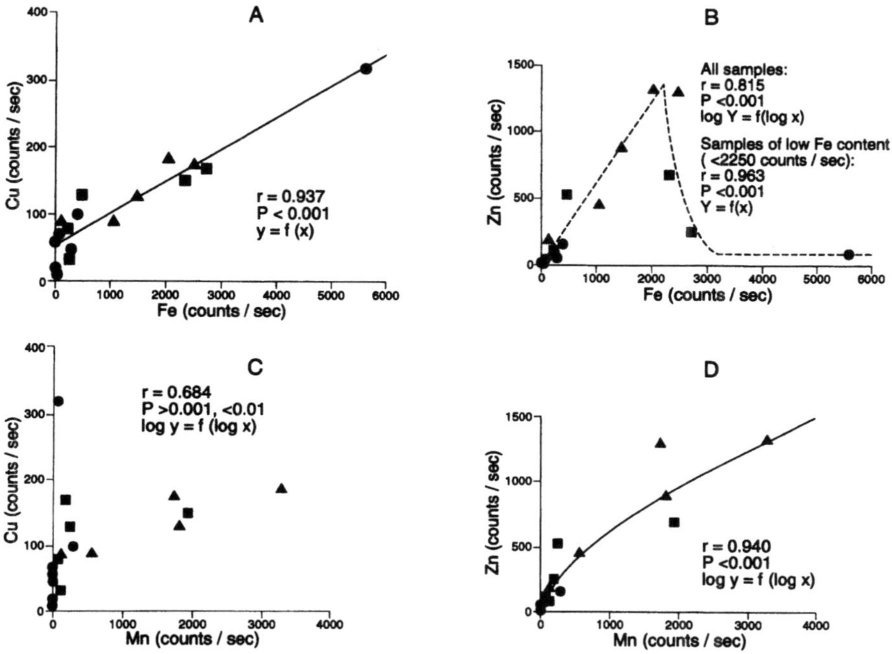

Fig. 5.15 A–D. Relationships between metals bound to remains of crustacean exoskeletons in plankton samples from epilimnion waters of three lakes polluted by a base-metal mining and smelting complex at Flin Flon, Manitoba (Canada). **A** Cu and Fe; **B** Zn and Fe; **C** Cu and Mn; **D** Zn and Mn. The biogenic particles in each sample were analysed individually by energy dispersive X-ray microanalysis in conjunction with electron microscopy. In the plots the *symbols* identify the lakes. (From Jackson and Bistricki 1995; republished with the permission of Elsevier Science-NL)

Price 1970, 1977; Förstner and Patchineelam 1980; Khalid 1980; Stiff et al. 1982; Osaki et al. 1990; Jackson and Bistricki 1995; Fig. 5.15). If the literature cited above is any indication, Mn oxides are usually more effective sorbents than Fe and Al oxides. Such generalisations, however, should be viewed with caution, as the sorption capacity of an oxide is a function of crystallinity, particle size, and specific surface as well as elemental composition. It has been observed repeatedly that the sorption capacity of any oxide is critically dependent on whether the oxide is "amorphous" or crystalline. Owing to their higher specific surface, freshly precipitated amorphous oxide gels invariably have much higher sorption capacities than the well crystallised phases that eventually form from these gels during ageing (Lee 1975; Rickard and Nriagu 1978; Pickering 1980b; Kester et al. 1986; Swift and McLaren 1991; Bowell 1994); indeed, crystallisation

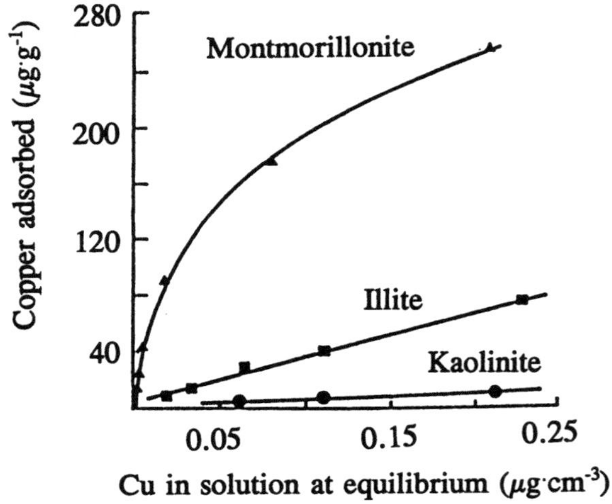

Fig. 5.16. Curves representing adsorption of Cu by montmorillonite, illite, and kaolinite. (From McLaren et al. 1981; republished with the permission of Blackwell Science Ltd. and R.G. McLaren)

may be accompanied by release of sorbed or coprecipitated metals (Waslenchuk 1975), presumably because they cannot be fitted into the crystal structure. Metals whose ionic radii are unsuitable for isomorphous substitution in the oxide are probably among the ones most likely to be excluded.

Also note that differences in the sorption capacities of Mn, Fe, and Al oxides vary with physicochemical conditions. Manceau et al. (1992) reported that below a critical pH the percentage of dissolved Pb(II) sorbed by Mn, Fe, and Al oxides varied in the order Mn $\gg$ Fe $>$ Al, implying that the properties of the Mn oxide were most conducive to the surface complexing of Pb^{2+} ions. In contrast, the three oxides were equally efficient sorbents at higher pH values, suggesting indiscriminate deposition of complex hydrolysates on any available surface. Coatings of organic matter, phosphate, and other impurities may also be important because, besides masking the surfaces to which they are sorbed, they affect the physical properties of oxides. Organic matter may either block the precipitation of Fe oxide (Schwertmann et al. 1986; Sato 1989) or promote it (Liang et al. 1993), and organic compounds (e.g. certain carboxylic acids) may retard the crystallisation of Fe oxide gels and clay minerals and affect their crystal structures, preserving unstable configurations indefinitely (Schwertmann et al. 1986; Jackson 1977, 1993a). Thus, organic substances, besides binding metals directly, may influence the metal-binding capability of an oxide or clay mineral.

Like oxides, clay minerals differ amongst themselves in their sorption capacities (Fig. 5.16). The cation exchange capacities of montmorillonite and vermiculite exceed those of other common clay minerals (such as kaolinite, illite, and chlorite) by an order of magnitude (Grim 1968). Therefore, if coulombic binding by cation exchange sites on 001 faces were the principal mechanism of heavy metal sorption, montmorillonite and vermiculite would consistently sorb these metals much more efficiently than any other clay mineral. Among clay minerals, however, ability to sorb heavy metals does not necessarily correlate well with cation exchange capacity. This is not surprising, for, as we have seen, surface complexation, which, in clay minerals, is limited to the edge faces, is the dominant mechanism for the sorption of heavy metals in nature, and a number of variables affect the outcome. This may help to explain why, as we will see, cation exchange capacity is not a reliable predictor of a clay mineral's sorption capacity for heavy metals, and why experiments by different workers have yielded inconsistent data on the relative affinities of clay minerals for heavy metals.

Different investigators working independently under various experimental conditions (e.g. over a range of pH values) have agreed that the sorption capacities of clay minerals for Hg(II) decrease in the order illite > montmorillonite > kaolinite (Reimers and Krenkel 1974; Andersson 1979) (Fig. 5.10). But a study of the specific sorption of Cu(II) by clay in dilute $CaCl_2$ solution at pH 5.5–6.0 showed that the order of preference varied with dissolved Cu concentration (Swift and McLaren 1991); at high Cu levels (up to 10 μg/ml) the order was illite > montmorillonite ≫ kaolinite, but at low levels (≤ 0.25 μg/ml) it was montmorillonite > illite > kaolinite, which is also the order of decreasing cation exchange capacity (Grim 1968) and decreasing "Cu exchange capacity" (Gomer et al. 1992). Rickard and Nriagu (1978) reported that the "intrinsic selectivity coefficients" for the exchange of Pb^{2+} for Ca^{2+} by clay varied in the order kaolinite > illite > montmorillonite; but Srivastava et al. (1989) found that the Pb^{2+} and Cd^{2+} sorption capacities of clay pretreated with acid and H_2O_2 decreased in the order montmorillonite > kaolinite. Similarly, Metwally et al. (1993) demonstrated that uptake of Zn by clay varied in the order montmorillonite ≫ kaolinite at pH 4.5–7.0, a result which they attributed (possibly on insufficient grounds) to the high cation exchange capacity of montmorillonite (while drawing the seemingly contradictory, but more plausible, inference that the Zn was bound by "chemisorption"). It may well be the higher specific surface of montmorillonite (a consequence of its smaller particle size) rather than its higher cation exchange capacity that accounts for the observed result. The sorption capacity of a clay mineral for a heavy metal is probably a complex function of the composition and structure of the mineral, the properties of the metal species, and various environmental parameters, and is therefore difficult to interpret and predict.

5.4.1.8
Surface Coatings

In their native state, clay crystals, along with silt and sand grains, are commonly coated to a greater or lesser extent with Fe and Mn oxides (Jenne 1968; Lee 1975; Waslenchuk 1975; Förstner and Patchineelam 1976, 1980; Gibbs 1977; Harmsen 1977; Andersson 1979; Beneš and Havlík 1979; Siegel and Siegel 1979; Pickering 1980b; Richter and Theis 1980; Snodgrass 1980; Kinniburgh and Jackson 1981; Lion et al. 1982; Boust and Joron 1986; Kester et al. 1986; Calder 1988; Jackson 1989, 1993b; Swift and McLaren 1991; Cowan et al. 1992; Zachara et al. 1992). Oxides may be deposited on the surfaces of other oxides as well; for instance, a layer of MnOOH could be precipitated on top of an FeOOH deposit following an increase in the ambient dissolved O_2 level. Oxide coatings are also deposited on exposed surfaces of the hard parts of marine and freshwater plankton and aquatic insects (Fowler 1977; Dunlap et al. 1983; Dunlap and Walne 1985, 1987; Barnes et al. 1986; Hare et al. 1991; Cain et al. 1992; Hare 1992; Jackson and Bistricki 1995) and on bacterial cells (Urrutia and Beveridge 1993). In addition, clay and oxide minerals, including oxide films on clay, may be coated with humic substances (Pickering 1980a, b; Lion et al. 1982; Hayes and Himes 1986; Kester et al. 1986; Goldberg et al. 1988; Förstner 1991; Swift and McLaren 1991; Krishnamurthy 1992; Marsi 1992; McKnight et al. 1992; Warren and Zimmerman 1994). Because oxides and humic matter are efficient scavengers of heavy metals and have much higher sorption capacities for these metals than clay minerals do, it is probable that the apparent sorption of heavy metals by clay, and other phenomena involving the surfaces of clay crystals, are partially, if not principally, due to these coatings (Siegel and Siegel 1979; Kester et al. 1986; Stotzky 1986; Jackson 1989, 1995). But effects of coatings are variable and complex (Pickering 1980b). Humic acid coatings on Fe oxide may either enhance trace element sorption (Tam et al. 1995) or decrease it (Pickering 1980a). Removal of oxide coatings has been found to increase the exchange capacity of clay while depressing the sorption of heavy metals such as Zn, suggesting that oxides obstruct exchange sites of clay but increase the number of specific sorption sites, although other factors such as the clumping of clays by oxides could also be involved (Pickering 1980b). However, Jackson (1989) observed that removal of oxide coatings from montmorillonite and kaolinite by reductive solvent extraction enhanced the sorption of Hg(II) by the clay; similarly, Ward and Bassett (1991) reported that removal of organic and oxide coatings from kaolinite and illite increased the sorption capacities of the clay minerals for Cr(VI), the change being especially marked in the case of illite. These results suggest that the surface impurities were blocking edge sites with strong affinities for the metal (or, perhaps, had caused clumping of the clay particles).

Other surface impurities on mineral particles include Al and phosphate species (Grim 1968; Sato 1989). Al may block the sorption of heavy metals (e.g. sorption of Cu(II) by kaolinite) (McBride 1978), and it clogs the exchange sites of clay (Grim 1968). This may result from strong fixation of Al^{3+} ions at sorption sites owing to their high charge/radius ratio. Phosphates, however, have complex and seemingly contradictory effects. According to experiments performed by different workers, sorbed phosphate promotes sorption of metals by allophane, goethite, and lepidocrocite (probably by increasing the number of negatively charged sites and by creating bridges linking the metal to the mineral surface) but has little effect on the sorption of Zn by montmorillonite and supresses the sorption of certain metals (e.g. Cu(II)) by acid clay, alkali clay, and $Al(OH)_3$, perhaps by blocking sorption sites (Bolland et al. 1977; Madrid et al. 1991; Schmitt and Sticher 1991; Du 1992).

In summary, coatings and other impurities on mineral surfaces are virtually ubiquitous, and they may radically alter the surface properties of the particles; consequently, the importance of their effects on the sorption and desorption of trace elements by these particles can scarcely be overemphasised. Surface coatings promote sorption if they are more effective sorbents than the particles on which they were deposited; but they may hinder sorption by blocking sorption sites on these particles (or possibly by promoting aggregation of mineral particles). Whether a surface deposit has the one effect or the other may depend on a combination of circumstances, and both effects could occur simultaneously; thus, the *net* result may be hard to predict.

Although natural oxide and organic coatings on clay crystals are widely regarded as important and are frequently mentioned in the literature, their role has seldom been assessed or even considered in research on interactions between heavy metals and clay minerals. A survey of the literature revealed that hardly any of the numerous experiments on clay-metal interactions which have been carried out by different workers over the past two decades included comparison of the effects of clay minerals before and after removal of coatings (Lewis et al. 1972; McBride and Mortland 1974; Payne and Pickering 1975; Sabet et al. 1975; García-Miragaya and Page 1977; Newton et al. 1976; Griffin et al. 1977; Babich and Stotzky 1977a; Farrah and Pickering 1978; McBride 1978; McLaren et al. 1981; Hamed et al. 1983; Debosz et al. 1985; Wang et al. 1985; Collins and Stotzky 1992; Sikalidis et al. 1989; Flemming et al. 1990; Nielsen 1990; Schulthess and Huang 1990; Brause et al. 1991; Davison et al. 1991; Helios-

Ryicka and Kyziol 1991; Madrid et al. 1991; Puls et al. 1991; Sharma et al. 1991; Stockmeyer and Kruse 1991; Wagner and Czurda 1991; Warith and Yong 1991; Zachara et al. 1991; Fujiyoshi et al. 1992; Gagnon et al. 1992; Gomer et al. 1992; Gupta et al. 1992; Hunter and James 1992; Kozar et al. 1992; Marsi 1992; Mosser et al. 1992; Sun Kou et al. 1992; Al-Sabri et al. 1993; Majone et al. 1993; Metwally et al. 1993; Stadler and Schindler 1993a; Baham and Sposito 1994). The results, though undoubtedly valuable in other respects, are somewhat ambiguous, because there is no way of telling to what extent they represent effects of coatings as distinct from effects of the clay minerals as such (McBride 1991a; also see Schulthess and Huang 1990, 1991). The few workers who have performed comparative studies designed to yield information on the role of coatings reported that clay minerals whose surfaces had been stripped of oxide deposits behaved differently than the same minerals with their natural coatings left intact, the differences ranging from neglible in some instances to very striking (and even critically important from the standpoint of microbial ecology and metal speciation) in others (Jackson 1989; Ward and Bassett 1991). Nevertheless, it is at least conceivable that clay surfaces are somehow altered by the chemical treatments required to strip off the coatings. This poses a difficult dilemma which remains to be resolved.

5.4.1.9
Selective Sorption of Trace Elements

Trace elements vary greatly in their relative affinities for sorption sites of clay minerals and oxides, the order of increasing preference being a complex function of the properties and speciation reactions of the elements, the structure and composition of the mineral, characteristics of individual sorption sites, environmental variables, and concentrations of competing trace elements. When in contact with water containing equal concentrations of various dissolved metal cations, the 001 faces of clay preferentially sorb the cations of highest charge and (among cations of the same charge) those that are largest and most weakly hydrated (Grim 1968; McBride 1976, 1980, 1991b). Given a mixture of dissolved cations of similar size and degree of hydration, these sorption sites do not discriminate between divalent heavy metals and alkaline-earth metals, mass action being the process that controls cation exchange (Pickering 1980a; Schmitt and Sticher 1991; Swift and McLaren 1991; see above). The unifying principle underlying the relationships between cation properties and the order of preference is expressed by the Born equation, which states that a cation's hydration energy (comprising enthalpy and entropy terms) is directly proportional to the square of the cation's charge and inversely proportional to its radius (Phillips and Williams 1965), implying that charge is the dominant factor. But the replaceability of one cation by another is also influenced by the nature of the associated anion (Grim 1968); and, in view of its effect on the solubility of salts and the stability of complexes, mutual polarisation of the anion and cation could be involved (Douglas and McDaniel 1965), as in the stabilisation of Hg in solution by the formation of Cl^- complexes, preventing sorption of the Hg.

The specific sorption sites of oxides and edge faces of clay show a much higher degree of selectivity than do the cation exchange sites on 001 faces of clay, and the outcome of competition between different metals for the ligands of these binding sites is determined by a larger, more complex assortment of factors (Harmsen 1977; Rickard and Nriagu 1978; McBride 1991b; see above). Divalent cations of heavy metals (e.g. Group IB and IIB metals) are sorbed preferentially with respect to alkaline-earth cations (Harmsen 1977; Madrid et al. 1991), and the relative affinities of the heavy metals for the sorption sites depend on a wide range of metal properties, along with specific characteristics of the mineral and the associated aqueous phase – factors which affect the stability of the partly covalent bonds formed between metal species and surface ligands of the mineral and determine the outcome of competition between these ligands and the ligands of the solution (Hingston et al. 1968; Leckie and James 1974; Forbes et al. 1976; Kinniburgh et al. 1976; Harmsen 1977; Rickard and Nriagu 1978; Pickering 1980a,b; Kinniburgh and Jackson 1981; Bruemmer et al. 1986; Brümmer 1986; McBride 1991b; Schmitt and Sticher 1991; Schuster 1991; Swift and McLaren 1991). The properties of a heavy metal that affect its affinity for a specific sorption site include polarisability, ionisation potentials, electronegativity, electron affinity, enthalpy and entropy of hydration, hydrolysis constants, and ionic radius (or the radius of the hydrated ion) (Harmsen 1977) – in short, characteristics which determine the stability and degree of covalence of the metal–ligand bonds. Oxides generally show a preference for "softer" nontransition heavy metal cations [e.g. Pb(II)] over "harder" ones [e.g. Cd(II)], but they prefer "harder" transition metals over "softer" ones (McBride 1991b). Hg(II), however, is exceptional in this regard; though very "soft," it is less efficiently sorbed by goethite than divalent Cd, Zn, Cu, and Pb are, supposedly because the Hg–OH bond of the $HgOH^+$ ion has an anomalously high degree of covalent character, weakening the ion's affinity for the oxide surface (Schuster 1991). The selectivity of cation binding is also influenced by ambient pH, as the sorption of readily hydrolysed metal cations is highly sensitive to pH, and each metal has its own optimal pH range for sorption (see Sec. 5.4.1.4 and Fig. 5.8). Selectivity based on pH presumably occurs with anionic as well as cationic trace element species

(Hingston et al. 1968; Swift and McLaren 1991). Moreover, the structure and composition of the mineral in the local region of each sorption site contributes to the process of selection; thus, the stability of the complex formed between a cation and the surface ligands must depend in part on the configuration of the ligands and the nature of the cationic constituents of the mineral structure to which the ligands are attached (i.e. whether the sorbed cations are bonded to Fe–OH, Mn–OH, Al–OH, Si–OH, or other groups or to some combination of groups). Finally, ionic radius probably plays a decisive part in determining which metal cations can be absorbed into the interior of a mineral and which ones are confined to the surface (Harmsen 1977; Förstner 1991). Sufficiently small cations may penetrate the hexagonal cavities of clay crystals and other openings in mineral structures (Harmsen 1977); and sorbed cations similar in size to a mineral's lattice cations could replace the latter by isomorphous substitution (diadochy) during slow infiltration of the mineral or during extension of the mineral structure, as in the growth of oxide deposits and edge faces of clay by accretion. For example, Cr(III) is able to substitute for Fe(III) in Fe oxides (Trolard et al. 1995). It is also important to note that desorption of metals from specific sorption sites is just as selective as the sorption process (Swift and McLaren 1991).

The laws governing the selective uptake and release of heavy metals by colloidal minerals are so numerous and diverse that they probably cannot be reduced to a universally applicable predictive formula (Swift and McLaren 1991). Although certain general tendencies can be discerned, there are deviations from them because so many different factors contribute to the net result. A few examples taken from the literature will suffice to illustrate these points. There have been several reported instances of positive correlation between the affinities of Fe oxides and other minerals for different heavy metal cations and the hydrolysis constants of the cations (Forbes et al. 1976; Bruemmer et al. 1986; Brümmer 1986; Sato 1989; Schmitt and Sticher 1991). At least two independent investigations of the uptake of divalent heavy metals by goethite have confirmed this relationship. Thus, the order of decreasing relative affinities of metals for goethite has been reported as Cu > Pb > Zn > Co > Cd by Forbes et al. (1976) and as Cu > Pb > Zn > Co > Ni > Mn by Schmitt and Sticher (1991). Moreover, Pb is sorbed more efficiently than Cd and Zn by Fe oxide encrustations on the surfaces of aquatic insects (Hare et al. 1991). These findings suggest that, as a general rule, specific sorption sites tend to bind hydrolysed metal cations preferentially. On the other hand, a tendency of the strength of metal binding to increase with ionic radius causes deviation from this rule (Bruemmer et al. 1986; Brümmer 1986), and there are deviations due to other factors too, including the composition and structure of the sorbent. Accordingly, experiments performed by Kinniburgh et al. (1976) demonstrated distinctly different relative affinities for Fe and Al oxide gels:

Fe oxide: Pb > Cu > Zn > Ni > Cd > Co;
Al oxide: Cu > Pb > Zn > Ni > Co > Cd.

Similarly, Pickering (1979) recorded the following affinity series for freshly precipitated Fe, Al, and Mn oxides:

Fe oxide: Pb > Cu > Zn > Ni > Cd > Co > Sr > Mg;
Al oxide: Cu > Pb > Zn > Ni > Co > Cd > Mg > Sr;
Mn oxide: Cu > Co > Mn > Zn > Ni.

Furthermore, experiments with various synthetic Fe and Mn oxides showed that the affinity of Pb for Mn oxide was up to 40 times greater than its affinity for Fe oxide (McKenzie 1980). By the same token, separate Fe and Mn oxide phases coexisting in aquatic environments differ in their trace element preferences and may even discriminate between different species of the same element (e.g. inorganic Hg and methyl Hg); consequently, the oxides differ considerably in trace element composition (Burns and Fuerstenau 1966; Calvert and Price 1970, 1977; Khalid 1980; Jackson et al. 1982; Stiff et al. 1982; Jackson 1988a; Osaki et al. 1990; Jackson and Bistricki 1995; Tessier et al. 1996). The results of some studies are in agreement about the relative affinities of metals for coexisting Fe and Mn oxides, but there is also disagreement. For instance, four independent studies of different systems have revealed preferential binding of Zn by the Mn oxide phase (Burns and Fuerstenau 1966; Calvert and Price 1977; Osaki et al. 1990; Jackson and Bistricki 1995; Fig. 5.15), but a fifth study showed selection of Zn by the Fe oxide phase (Calvert and Price 1970); similarly, three studies agreed that Cu was concentrated preferentially in Mn oxide (Burns and Fuerstenau 1966; Calvert and Price 1977; Stiff et al. 1982), whereas a fourth showed selective accumulation of Cu by Fe oxide (Jackson and Bistricki 1995; Fig. 5.15). These apparent inconsistencies show that separation of trace elements owing to selective accumulation by different binding agents is not altogether predictable, probably because the observed result is the net effect of many factors, such as properties of the mineral, ambient conditions, metal speciation and complexing, and microbial activities, which vary from one system to another.

As with oxides, clay minerals differ considerably among themselves in their relative affinities for metals, as illustrated by the following set of data for the sorption of divalent cations (Pickering 1980a,b):

montmorillonite: Ca > Pb > Cu > Mg > Cd > Zn;
illite: Pb > Cu > Zn > Ca > Cd > Mg;
kaolinite: Pb > Ca > Cu > Mg > Zn > Cd.

In a study mentioned by Siegel and Siegel (1979), sorption of divalent cations by montmorillonite yielded similar results except that Ca had a different place in the series with respect to Pb and Cu, the order of preference being Pb > Cu > Ca > Ba > Mg > Hg. The relative ease of cation replacement was, of course, the reverse of this series. For kaolinite, sorption energy decreased in the order Hg > Cu > Pb, but calcium replacement decreased in the order Pb > Cu > Hg, signifying that sorption *strength* was inversely related to sorption *capacity*. Thus, Hg was most strongly bound, but the quantity of Hg taken up was smallest.

Since humic matter is commonly bound to clay and oxide surfaces, the relative stabilities of different humic–metal complexes should be considered. Here, too, we find systematic variations but disagreement as well as agreement between the results of different studies. Thus, the relative stabilities of humic–metal complexes correlate with metal properties, but the relationships vary from one natural or experimental system to another, as illustrated by the following sequences representing decreasing stability of humic–metal complexes as reported by different workers:

Pickering (1980 a, b): Pb > Cu > Cd = Zn > Ca. (Stability correlates positively with electronegativity; Douglas and McDaniel 1965.)
Jackson et al. (1980 a): Hg > Co > Cr > Zn > Mn > Cs. (Stability shows a strong tendency to increase with electronegativity.)
Swift and McLaren (1991): $Fe^{3+} > Cu^{2+} > Zn^{2+} > Co^{2+} > Mn^{2+}$. (Stability increases with the sum of the relevant ionisation potentials; Lide 1992.)
Schnitzer and Khan (1972): $Pb^{2+} > Cu^{2+} > Ni^{2+} > Co^{2+} > Zn^{2+} > Cd^{2+} > Fe^{2+} > Mn^{2+} > Mg^{2+}$. (This array tends to parallel the Irving-Williams series, or "natural order of stability"; Douglas and McDaniel 1965.)

Also note that the proportion of the humic–metal fraction that is bound to the highest molecular weight fraction of the humic matter has been found to increase with the electronegativity of the metal (Jackson et al. 1980 a).

In summary, data for clay minerals and oxides are variable and may be ambiguous and hard to interpret. To begin with, they represent the net result of reactions occurring at binding sites with different properties, and the reactions vary with environmental conditions. Besides, coatings and other impurities on mineral surfaces affect metal uptake but are almost never taken into account. Since all these factors enter into the equation and may vary independently, it is not surprising that there are apparent contradictions between experimental results reported by different investigators.

5.4.2
Coprecipitation and Solubilisation

There is no essential difference between the sorption and coprecipitation of metals by oxides except that in coprecipitation the sorbed metals are assimilated into a growing mass of oxide as it precipitates – usually, in the case of Fe and Mn oxides, at the boundary between a reducing (or anoxic or O_2-poor) environment and an oxidising (or O_2-rich) one, or where extremely acidic waters charged with dissolved Fe are being neutralised (Lee 1975; Karlsson et al. 1987). Oxide stability depends on Eh, pH, and complexing agents. The lower the pH, the higher the Eh must be for precipitation of the oxide to begin, although complexing of Fe or Mn may interfere, even if the Eh and pH favour precipitation. With rising or falling Eh at constant pH, Mn oxide starts to precipitate or dissolve, respectively, at a higher Eh than Fe oxide does. As Mn and Fe oxides differ in their relative affinities for metals, this implies selective scavenging and release of metals. In nature, precipitation of Fe and Mn oxides is often mediated by bacteria (e.g. *Ferrobacillus*). Metals occluded in masses of oxide owing to coprecipitation probably cannot be released back into solution unless a shift in environmental conditions causes solubilisation of the oxide.

5.4.3
Flocculation and Peptisation

When clay minerals and oxides are dispersed, or peptised, in water, each particle is surrounded by an electrical double layer formed by the surface charge and a diffuse cloud of oppositely charged ions. As long as the particles have the same net surface charge, they are kept apart by mutual electrostatic repulsion. If exposed to dissolved salts, as when clay is transported to the sea by a river, the dispersed particles are liable to flocculate owing to weakening of the surface charge, with shrinkage of the electrical double layer and lowering of the zeta potential, allowing the particles to come close enough to each other to be stuck together by van der Waals' forces (van Olphen 1963; Andersson 1977; Harmsen 1977; Duarte et al. 1991). The higher the valence of the ions of opposite charge to the colloid, the more efficient are those ions in flocculating the colloid; thus, as a rule, the most efficient flocculating agents for clay are polyvalent cations (although ions of the same valence differ somewhat among themselves in their flocculating power) (van Olphen 1963). Organic substances, including humic matter and tannates, may protect clay against flocculation by salts, keeping it in a dispersed state (Whitehouse et al. 1960; van Olphen 1963; Narkis et al. 1968); alternatively, organic compounds may flocculate clay or "sensitise" it (i. e. promote flocculation of the clay by salts), or may alternately flocculate and peptise the clay as the concentration of the organic substance in solution varies (van Olphen 1963). Specific sorption of a metal cation, or a change in the ambient pH (which determines – and may reverse – the net charge of an oxide particle or the edge of a clay crystal), may also promote flocculation by causing the colloid to attain or approach its "zero point of charge" (at which there is no *net* charge) (Harmsen 1977) or point of zero zeta potential (Leckie and James 1974). However, the effects of pH are complex. Alkali may either peptise or flocculate clay; and acids could promote flocculation by creating a net positive charge on the edge faces, causing edge faces of some particles to bind the negatively charged 001 faces of others (van Olphen 1963).

5.4.4
Minerals as Scavengers, Carriers, and Secondary Sources of Trace Elements in Natural Environments

5.4.4.1
The Binding and Transport of Metals by Colloidal Particles

When dissolved heavy metals are introduced into aquatic environments by natural processes or discharge of wastewater, they are rapidly and more or less efficiently taken up by fine-grained suspended particles and are concentrated in the clay and silt fractions of bottom sediments; they are also accumulated by organisms. Metals bound to sediments may be released back into solution to a greater or lesser extent depending on the circumstances; consequently, fine-grained sediments are secondary sources as well as sinks for heavy metals. These partitioning processes are functions of environmental parameters (e.g. pH, Eh, and salinity), the nature and abundance of the particles, the properties of the metals, and the activities of microbes and other organisms (which largely control environmental conditions and geochemical phenomena in the biosphere). The same principles, of course, apply to soil and subsoil environments.

Because of their large specific surfaces and high sorption capacities, the clay-sized fractions of sediments and soils selectively accumulate trace elements. Coarser particles (e.g. sand-sized fractions) are relatively poor in trace elements, except in rare circumstances such as the occurrence of coarse organic detritus with a high affinity for trace elements

(Kudo et al. 1977; Jackson and Woychuk 1980a, b; Jackson et al. 1982). Elements bound to carrier particles are subject to redistribution by the ordinary processes of erosion, transport, sorting, and deposition (erosion of soil by wind and runoff, translocation in soil, fluvial transport of particles, the settling out of particles in bodies of standing water and in slow-moving river water, resuspension of sediments by bottom scour, and so on). The importance and universality of the scavenging and redistribution of trace elements by clay-sized minerals and other colloids in the biogeochemical cycling of trace elements have been attested by an immense body of literature (de Groot et al. 1971; Cranston and Buckley 1972; Gavis and Ferguson 1972; John 1972; Thomas 1972; Bothner and Carpenter 1973; Leckie and James 1974; Reimers and Krenkel 1974; de Groot and Allersma 1975; Shuman 1975; Waslenchuk 1975; Förstner and Patchineelam 1976, 1980; Gibbs 1977; Harmsen 1977; Kudo et al. 1977; Turekian 1977; Hogg et al. 1978; Rickard and Nriagu 1978; Salomons and de Groot 1978; Andersson 1979; Beneš and Havlík 1979; Boyle 1979; Fitzgerald 1979; Leckie and Davis 1979; Pickering 1979, 1980a,b; Thornton 1979; Windom and Kendall 1979; Förstner 1980, 1991; Snodgrass 1980; Young et al. 1980; Jackson et al. 1982; Rogers et al. 1984; Jackson 1986, 1993b; Livens et al. 1986; Salomons and Baccini 1986; Karlsson et al. 1987; Belzile and Tessier 1990; Nielsen 1990; van der Weijden 1990; Morgan and Stumm 1991; Ohnesorge and Wilhelm 1991; Scheinberg 1991; Stoeppler 1991; Belamie and Gouy 1992; Honeyman and Santschi 1992; Peters and Shem 1992).

Published estimates of the efficiency of metal scavenging by particulate matter vary widely. According to a review by Fitzgerald (1979), Hg in rivers and lakes is 10–90% particulate and rises to as much as 98% in estuarine environments but is mostly "dissolved" (< 0.45 μm in diameter) in coastal seawater (though much of the "dissolved" fraction is admittedly "associated with organic material," including low molecular weight fulvic acids). However, the accuracy and even the meaningfulness of such data, especially the lower values, are open to doubt, because the conventional techniques for separating dissolved and particulate fractions of water are too crude to be useful except for rough preliminary estimates and gross semi-quantitative comparisons. The usual method of fractionation is to pass water samples through 0.45 μm membrane filters, but this does not lead to a reliable estimate of the proportions of the metals in the solution and particulate phases. A major disadvantage of the method is that many dispersed particles are small enough to pass through membrane filters (the smallest dimensions of colloidal particles in lake water falling in the range 0.001–1.0 μm) (Morel and Gschwend 1987; Leppard et al. 1989; Buffle et al. 1992; Honeyman and Santschi 1992; Horowitz et al. 1996). The result is that a large but unknown proportion of the metals in the operationally defined "solution" phase may actually be in suspension. Furthermore, any kind of filtration process - not only membrane filtration but also ultrafiltration and dialysis, which have been tried out to a more limited extent - may alter the sample and create artefacts (e.g. by contamination of the water, or sorption of solutes, by the filter, or by induced coagulation of sample constituents), and the performance of the filter may be changed by interaction of the filter with the sample (e.g. by the clogging of pores) (Buffle et al. 1992). Besides filtration, field flow fractionation and hydrodynamic chromatography have been attempted, but these methods have met with only limited success (Buffle et al. 1992).

Next to filtration, centrifugation has been the most popular method for separating dissolved and particulate matter, but it too is subject to limitations and problems. Centrifugation in the field using specially designed portable equipment has been employed, especially as a means of collecting bulk quantities of suspended fluvial sediment for analysis; but, although the efficiency of particle recovery by centrifugation is often >90% for fine mineral particles (which predominate during episodes of high discharge, such as the spring flood in temperate regions), the efficiency is much lower (in some cases <50%) for fine organic particles (which predominate at times of low flow rates) (Ongley and Blachford 1982; see also Jackson et al. 1982). Therefore, the operationally defined "dissolved" fractions of metals in natural waters may, in fact, be largely bound to colloidal particles. Perhac (1972) used ultracentrifugation to distinguish between dissolved, colloidal, and coarser particulate metals in stream water, but the colloidal fraction as he defined it (0.01 - 0.15 µm) constitutes only a portion of the generally recognised colloid size range (0.001 - 1 µm), which extends far into the realm of his "solution" phase. Besides, the high rotor speeds used for ultracentrifugation (40 000 rpm) could have caused disintegration of certain particles, creating artefacts and leading to overestimation of the "dissolved" component. Therefore, Perhac's conclusion that the metals were mostly in solution should be viewed with scepticism.

This is not to say that the data for "dissolved" and "particulate" trace elements reported in the literature are worthless, but it does mean that we cannot accept them at face value and must be aware of their limitations. For in-depth investigation of the partitioning of metals between particulate and solution phases, it is necessary to apply advanced, refined techniques of sample preparation, electron microscopy, and energy dispersive X-ray microanalysis (Leppard et al. 1989, 1990). But these methods are so specialised, expensive, and time-consuming that they are not readily accessible to most investigators.

5.4.4.2
Partitioning of Trace Elements Among Different Solid Phases

The clay-sized fractions of soil, suspended matter, and bottom sediments are made up of different constituents bound to each other to form composite particles, or mixed together physically, in varying proportions. Heavy metals sorbed to this particulate matter are distributed among its various components in different proportions, which are determined by the nature and concentrations of the metals and metal-binding solid phases, and by the ambient conditions. In environments where there is appreciable free O_2, such as river water and soils, the main metal-binding agents are Fe and Mn oxides, clay minerals, organic matter (both humic and nonhumic), and carbonate minerals (FeS becoming a significant metal scavenger only in reducing environments where free H_2S is generated and free O_2 is absent, as in certain productive bodies of water and in waterlogged soils) (Feick et al. 1972; Thomas 1972; Waslenchuk 1975; Förstner and Patchineelam 1976, 1980; Andersson 1977; Gibbs 1977; Salomons and de Groot 1978; Beneš and Havlík 1979; Pickering 1979, 1980a, b; Snodgrass 1980; Jackson et al. 1982; Livens et al. 1986; Wilkins et al. 1986; Nielsen 1990; Förstner 1991; Schuster 1991; Modak et al. 1992; Jackson and Bistricki 1995). As would be expected from the experimental data discussed in Section 5.4.1.7, heavy metals are generally bound more effectively by oxides and organic matter than by clay minerals. However, there are deviations from this rule owing to overlap of the estimated ranges of the metal-binding capabilities of these three classes of colloids, as well as specific affinities of certain metals for particular colloids. In a study of metal partitioning in soil Andersson (1977) found that Pb and Cd were bound preferentially by organic matter, whereas Ni, Co, and Cr were associated with clay minerals, whilst Cu and Zn were divided between the clay and organic phases. To take another example, Abd-Elfattah and Wada (1981) reported that selectivity coefficients representing the relative preferences of different soil constituents for Cu^{2+}, Zn^{2+}, and Pb^{2+} with respect to Ca^{2+} decreased in the order Fe oxides > halloysite > allophane, imogolite > organic matter, kaolinite > montmorillonite. Thus, organic matter fell within the range of values for clay minerals. (It has to be borne in mind, though, that in both sets of results, as with most data of this kind, there is some ambiguity because coatings on the clay were not taken into account.) In any case, at low pH organic matter is a much more effective metal-binding agent than mineral colloids (Andersson 1977, 1979; Jackson and Woychuk 1980a,b; Schuster 1991; Tessier et al. 1996; Fig. 5.10).

Jackson (1979) has developed a unique application of regression analysis to the study of preferential associations of heavy metals with binding

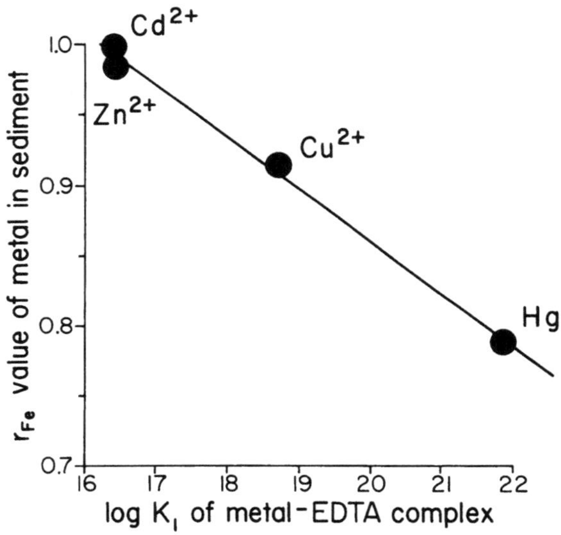

Fig. 5.17. Correlation coefficients (r_{Fe} values) representing the degree of association of various heavy metals with Fe (presumed to be mainly FeOOH) in well-oxygenated lake sediment, from Ball Lake (Wabigoon-English River system), Ontario, Canada (Fig. 5.18), plotted against the stability constants (log K_1 values) of the EDTA complexes of the metals (From Jackson 1979). Sources of log K_1 data: Sillén and Martell (1964, 1971). Values of regression parameters: $r = -0.997$; $P > 0.02, < 0.05$

agents in freshwater sediments. This involves the use of correlation coefficients (r-values) as estimates of the relative affinities of different metals for sediment components (e.g. Fe, organic C, and sulfide). The r-values (or ratios of r-values) are compared with published values for parameters that quantify specific properties of the metals (e.g. electronegativity), and significant correlations between them are then used to interpret the biogeochemical pathways of the metals. In one case involving clay-silt sediments from a well aerated lake, the affinities of Hg, Cu, Zn, and Cd for Fe (presumably in the form of FeOOH) gave a strong inverse correlation with the stability constants (log K_1 values) of EDTA complexes of the metals (Fig. 5.17); thus, the order of decreasing correlation with Fe was Cd, Zn > Cu > Hg. Log K_1, in turn, is a function of electronegativity. The observed relationship suggests that the metals which form the most stable and highly covalent bonds with chelators have the strongest tendency to be bound by humic matter in preference to Fe oxide in that particular lake (Jackson 1979).

As pointed out in Section 5.4.1.9, oxides and clay minerals differ among themselves in their metal-accumulating propensities. For instance, the physically separated but contiguous Fe and Mn oxide phases of marine and freshwater oxide deposits, including Mn-Fe nodules, differ from each other in trace element composition (Burns and Fuerstenau 1966; Calvert and Price 1970, 1977; Förstner and Patchineelam 1980; Khalid 1980; Stiff et al. 1982; Osaki et al. 1990; Lind and Anderson 1992; Jackson and Bistricki 1995; Fig. 5.15) and may even discriminate between different forms of the same element (Jackson et al. 1982 [see below]; Jackson 1988a). Similarly, clay minerals differ appreciably in their specific sorption capacities for heavy metals. Therefore, differences in the proportions of kaolinite, montmorillonite, illite, and other clay minerals in sediments and soils could affect the concentrations and spatial variations of metals in those materials (Reimers and Krenkel 1974; Rickard and Nriagu 1978; Srivastava et al. 1989; Swift and McLaren 1991; Metwally et al. 1993).

The distribution of metals among different coexisting colloidal phases varies with external conditions as well as the inherent properties of the colloids. Thus, the partitioning of Cd among the cation exchange sites and specific sorption sites of clay minerals and Fe and Al oxides in soil varies with the ionic strength and pH of the soil solution (Cowan et al. 1992; Zachara et al. 1992). In both soils and sediments, Fe oxides become enriched in metals (e.g. Eu) at the expense of humic acid at relatively high Eh and pH (Maes and Cremers 1986; Förstner 1991). In acidic soil, organic matter is the only effective binding agent for Hg, but near neutrality Fe oxide and clay are relatively important and may predominate (Andersson 1979; Schuster 1991; Fig. 5.10); similar results have been reported for metals in lake sediments (Tessier et al. 1996). In oxidised lake sediments, acidification favours accumulation of metals by organic matter and Fe oxides rather than Mn oxides, and binding by cation exchange sites is decreased, but when the pH is close to neutrality the reverse is true (Young and Harvey 1992). Oxidation–reduction reactions, as well as pH, control the formation and dissolution of Fe and Mn oxides, thereby regulating the binding and release, respectively, of heavy metals by these oxides. At a given pH, Fe and Mn oxides precipitate when dissolved Fe(II) and Mn(II) in anoxic or O_2-poor water come into contact with sufficient dissolved O_2 to cause microbially mediated or abiotic oxidation. As Mn(II) is stable at higher O_2 and Eh levels than Fe(II) is (Degens 1965; Förstner 1980), separate Fe and Mn oxide phases are often formed in different proportions depending on the oxidation–reduction conditions; and since Mn and Fe differ in their heavy metal affinities, this process affects the partitioning of metals (Lind and Anderson 1992; Jackson and Bistricki 1995; Fig. 5.15). On the other hand, if Fe and Mn oxides are subjected to reducing conditions, they tend to dissolve, releasing their bound metals, which may then be seized by sulfides and organic matter. In such circumstances, many heavy metals (e.g. Hg, Cu, Zn, and Cd) are preferentially bound by sulfides, but the metals appear to be partitioned between competing sulfide and organic phases (Jackson 1978, 1979).

5.4.4.3
Effects of Spatial and Temporal Changes in Environmental Conditions

If dissolved and suspended metals are transported from one environmental regime to another, or an interface or transition zone between two regimes arises, or conditions at a certain locality change over time, bound metals may be solubilised, carrier particles may flocculate, or dissolved metals may be immobilised by sorption or coprecipitation, resulting in redistribution of the metals among different compartments of the environment. Striking examples of this occur along environmental gradients (as in estuaries, rivers, and river–lake systems), at the boundary (or in the transition zone) between oxidising and reducing environments (as in a lake in which oxygenated water overlies anoxic hypolimnion water or anoxic bottom sediments), and in any environments subjected to progressive, periodic, or occasional alteration over time (e.g. long-term acidification or the cycle of seasonal changes in rivers and lakes).

Estuaries. Metals sorbed to carrier particles transported to the sea by a river undergo significant changes in form and distribution on entering the estuarine environment, in which river water grades into seawater. As the salinity and ionic strength of the water increase, many metal cations are desorbed to a greater or lesser extent, reportedly by the following mechanisms: (1) reaction of the metals with Cl^- ions and other simple inorganic anions in seawater to form soluble complexes; (2) competitive displacement of the metals from exchange sites by Mg^{2+}, Na^+, and other cations in seawater; (3) intensified microbial decomposition of suspended organic matter, resulting in release of bound metals and synthesis of secondary organic products that form soluble metal complexes; and (4) solubilisation of humic–metal complexes because of the mildly alkaline pH of seawater (Kharkar et al. 1968; de Groot et al. 1971; Leckie and James 1974; de Groot and Allersma 1975; Beneš and Havlík 1979; Leckie and Davis 1979; Förstner 1980; Khalid 1980; Pickering 1980a; Young et al. 1980; van der Weijden 1990; Kozar et al. 1992; Modak et al. 1992; Chen et al. 1995; Turner and Tyler 1997). As heavy metals are bound mainly by specific sorption, the formation of soluble complexes probably predominates over displacement by alkali and alkaline-earth cations (Leckie and James 1974; Wang et al. 1991). As for the rise in pH on approaching the sea, it is probably not large enough to cause more than a moderate increase in the solubility of humic matter.

Owing to differences in the metals' tendencies to form stable dissolved complexes and to be desorbed from particles, entry into the transition zone between the freshwater and marine domains leads to a change in the proportions of the metals in the particulate and solution phases. De Groot and Allersma (1975) reported that 78–95% of the particulate Cd, Hg, Cu, Zn, Pb, Cr, and As introduced into the Rhine River estuary was solubilised, whereas only 50% of the Ni was solubilised, and La, Sc, and Sm remained bound to particles. Hg has a greater tendency than certain other metals (e.g. Cu, Pb, Zn, and Co) to be mobilised as complexes because it forms stable complex ions (de Groot et al. 1971). Hg and Cd are among the metals most prone to solubilisation by Cl^- (Leckie and James 1974; de Groot and Allersma 1975; García-Miragaya and Page 1976; Newton et al. 1976; Khalid 1980; van der Weijden 1990; Schuster 1991; Kozar et al. 1992; Roy et al. 1993; Chen et al. 1995), and this process is not limited to estuaries: contamination of Hg-polluted freshwater environments with Cl^- from chlor-alkali plant effluents and road deicing salt leads to desorption of Hg from sediments (Feick et al. 1972; Jackson et al. 1982 [see below]).

Another important phenomenon responsible for the redistribution of metals in estuaries is flocculation of dispersed metal-bearing colloids (clay minerals, oxides, and humic matter) by the dissolved salts of seawater (Turekian 1977; Leckie and Davis 1979; Kester et al. 1986; van der Weijden 1990; Duarte et al. 1991). Flocs with high settling velocities are deposited on the bottom, carrying bound metals with them, while others are carried out to sea (Kester et al. 1986). As flocculation starts at relatively low salinities, sedimented flocs tend to accumulate near the upstream end of an estuary, causing the bottom sediments to become sinks for heavy metals (Kester et al. 1986; van der Weijden 1990). The clumping of colloidal particles to form flocs may interfere with the desorption of metals by decreasing the total surface area exposed to the ambient salt solution (Duarte et al. 1991).

Although these generalisations may be widely applicable to estuarine environments, seemingly contradictory observations have been reported. Despite an accumulation of mostly experimental evidence that Hg is highly prone to desorption from particles by Cl^- ions owing to the formation of stable water-soluble Cl^- complexes (Lockwood and Chen 1973; Leckie and James 1974; Reimers and Krenkel 1974; de Groot and Allersma 1975; Lee 1975; Newton et al. 1976; Kinniburgh and Jackson 1978; Beneš and Havlík 1979; Inoue and Munemori 1979; Jackson et al. 1982; Wang et al. 1985, 1991; Schuster 1991; Chen et al. 1995), apparent exceptions have been recorded. Bothner and Carpenter (1973) found that Hg strongly bound to suspended matter in the Columbia River (northwestern United States) was practically insoluble in seawater, probably owing to the formation of extremely stable complexes with ligands in certain sediment components (Beneš and Havlík 1979); and experiments by Duarte et al. (1991) showed that an increase in the ionic strength of the ambient solution resulted in a *decrease* in the release of Hg from colloidal sediment particles, probably because flocculation interfered with desorption. Furthermore, field observations supported by results of laboratory experiments suggest that apparent desorption of Hg from suspended matter in the Wyre River estuary (England) may be due, instead, to dilution of fine-grained, highly organic particles by coarser, less organic detritus (Rae and Aston 1982); an increase in Cl^- concentration paralleled the dilution effect, giving the false impression that the observed effect is due to desorption by Cl^- ions. Field studies often indicate that in estuaries Zn, too, has a stronger tendency to be sorbed than to be desorbed, even though experimental results predict desorption by salts (Hegeman et al. 1992); this paradox serves to remind us that it is not always possible to extrapolate results of laboratory experiments to complex natural environments. Furthermore, salts may have complex effects on the sorption and desorption of metals by sediments; thus, Mayer and Schick (1981) demonstrated that the uptake of Cr(VI) by estuarine sediment increased, peaked, and then decreased with increasing salinity, suggesting varying effects of salinity on metal speciation or on the sorption sites, or both. Other seemingly contradictory effects of salinity have been mentioned by Khalid (1980). In addition, sea salts may have opposite effects on different metals. Experiments on the sorption of Cu(II) by Mn oxide showed that both seawater and artificial Cl^- solutions promoted sorption instead of interfering with it (Zaitseva and Varentsov 1990). A possible reason for this result is the fact that Cl^- ions form less stable dissolved complexes with Cu than with certain other metals (e.g. Hg), and are therefore less capable of desorbing Cu (Schuster 1991). In the experiments, the Cl^- ions may have removed competing metals from some of the sorption sites on the Mn oxide, clearing the way for sorption of Cu. It may be mentioned, as well, that some dissolved ions in a complex salt solution like seawater could counteract the effects of others; at any rate, Hoins et al. (1993) showed that SO_4^{2-} ions foster the sorption of Cd by goethite, whereas Cl^- ions, as we have seen, interfere with the sorption of Cd by colloids. Finally, further complications may arise as a result of seasonal environmental changes, such as stagnation causing reducing conditions and immobilisation of metals by sulfide production in the summer followed by more vigorous circulation and aeration resulting in solubilisation during the winter (Holmes et al. 1974). In brief, the fate of a metal in an estuary is the net result of many different conditions and may well vary from one estuary to another in ways that are not altogether predictable (Leckie and Davis 1979).

River systems. Metals introduced into a river system are exposed to temporal and spatial variations in environmental conditions which affect metal speciation, the binding and release of metals by particulate matter and complexing agents, and metal transport and deposition. Temporal variations include seasonal and weather-related changes in the river's discharge, dissolved and suspended solids, bottom sediments, and biological activity, and changes due to the input of pollutants over the course of time, as well as gradual recovery following cessation of pollution. Spatial variations include changes in water and sediment composition, discharge, suspended load, and biological activity that take place along the length of the river. Any river is subject to natural changes (e.g. an increase in discharge) as it flows down its gradient, or when it flows through a lake, merges with another river, or flows from one geological or environmental domain into another; in addition, spatial changes occur when pollutants are introduced into a river and when control structures such as dams are installed. When metals are introduced into river water from a point source, the metal concentrations decrease progressively in the downstream direction owing to dilution. Dilution also occurs when a metal-rich river flows into a metal-poor river. Experiments performed on Columbia River water and artificial "river water" appear to lend credence to the idea that dilution results in partial desorption of heavy metals, including Hg and Cu, from suspended sediments (Bothner and Carpenter 1973; Boyle 1979). Nonetheless, in view of the strong surface complexing of such metals by colloids and the probability that available surface complexation sites in rivers vastly outnumber dissolved heavy metal ions, it seems doubtful whether dilution alone can cause an appreciable increase in the ratio of dissolved metals to suspended ones in a river. A substantial shift in environmental conditions is more likely to produce a change of that kind.

A brief summary of selected results of the writer's research on the Hg-polluted Wabigoon River and Moose Jaw River–Qu'Appelle River systems of Canada will serve to illustrate these general principles. The Wabigoon River system (including two riverine lakes and a small reservoir) in the Boreal forest zone of Ontario (Fig. 5.18) provides a dramatic example of trends in heavy metal speciation, binding, release and bio-availability caused by a progressive change in physicochemical conditions and related microbial activities with increasing distance downstream from a source of pollution (Jackson and Woychuk 1980a, b, 1981; Jackson et al. 1982; Parks et al. 1991). Large quantities of Hg and coarse wood frag-

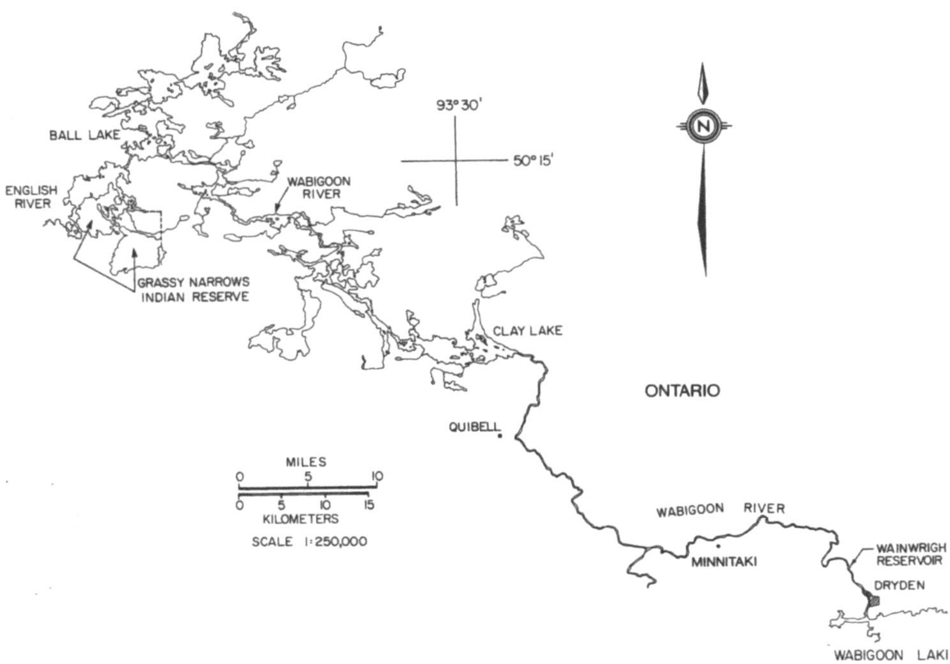

Fig. 5.18. Map of the Wabigoon River system, Ontario, Canada. The river flows northwest from its source (Wabigoon Lake) to its confluence with the English River (at Ball Lake). (From Jackson and Woychuk 1980a, b)

ments, along with NaCl and other pollutants, were introduced into the river from a chlor-alkali plant and pulp-and-paper mill at the town of Dryden over a period of years, resulting in a pronounced gradient in the composition of the surficial sediments going downstream from the industrial complex. High concentrations of both Hg and rotting wood chips near the source of pollution give way to progressively lower levels with distance downstream from it, the organic particles grading into natural clay-silt sediment (Fig. 5.19A). Jackson and Woychuk (1980a, b, 1981) and Jackson et al. (1982) showed that as the sediment graded from organic debris to clay-silt mud, the inorganic Hg in the sediment became more loosely bound (as might be expected) and hence more readily solubilised by Cl^- ions in river water (Fig. 5.20) and by mild reagents such as 0.5 M $CaCl_2$, which were used to extract the more weakly sorbed bioavailable (biochemically reactive) inorganic Hg fractions (Figs. 5.19B and 5.21). Microbial production of methyl Hg from inorganic Hg and the percentage of Hg in the form of methyl Hg (Jackson et al. 1982), as well as the Hg content of certain fish species (Jackson and Woychuk 1980a, b;

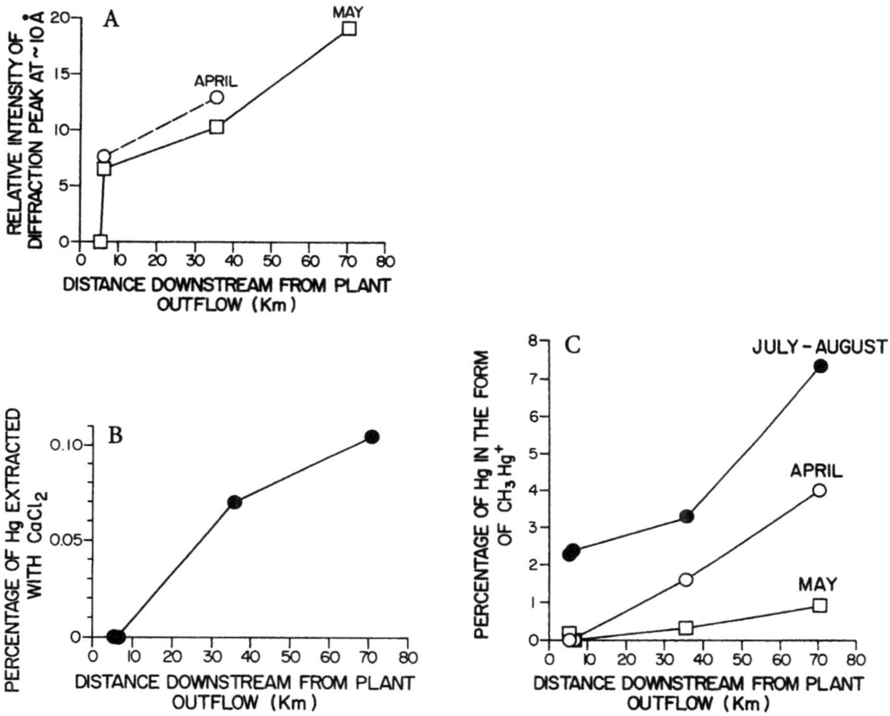

Fig. 5.19 A–C. Plots showing variations in the composition of suspended particles and water in the Wabigoon River (Fig. 5.18) as functions of distance downstream from the sources of Hg pollution at different seasons in 1979. **A** Illite content of suspended matter in April and May (during the spring flood); **B** percentage of the total suspended Hg that was extractable with 0.5 M $CaCl_2$ in July and August; **C** percentage of the total Hg content of unfiltered water that was in the form of methyl Hg (CH_3Hg^+) in April, May and July–August. Note: the spatial distribution of quartz was similar to that of illite. (Jackson et al. 1982, and unpubl. data; **A** was republished with the permission of the Editor-in-Chief of Hydrobiologia)

Parks et al. 1991), also increased in the downstream direction (Fig. 5.19 C) even though the total Hg concentration (mostly inorganic Hg, much of which is not readily bio-available) *decreased* sharply (Fig. 5.21). These observations indicate a progressive downstream increase in the bio-availability of Hg in the Wabigoon River. The trend could be explained by any of three causes (or a combination of them): (1) Hg is bound more loosely by clay and silt particles than by the organic detritus; (2) the clay-silt mud is more finely divided (i.e. has a greater exposed surface) than the organic detritus; and (3) conditions for the growth and activities of methylating microbes become more favourable farther away from the

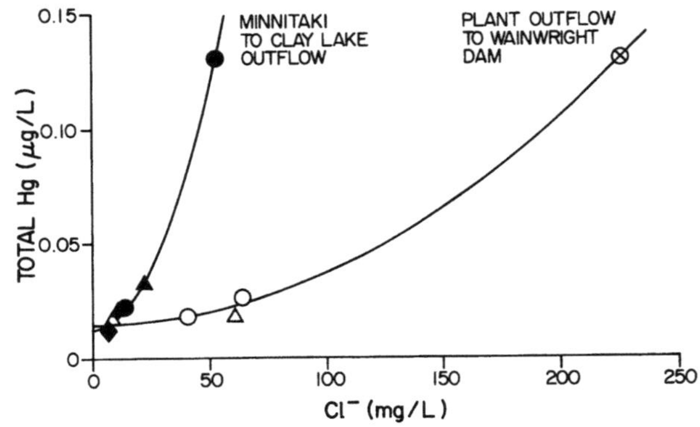

Fig. 5.20. Relationship between total Hg and Cl^- in centrifuged water collected from two different regions of the Wabigoon River system (Fig. 5.18) during the summer of 1979. The sampling areas extend (1) from the chlor-alkali plant at Dryden, Ontario to the outflow of Wainwright Reservoir (*white symbols*; bottom sediments blanketed with decomposing wood chips); and (2) from Minnitaki to the outflow of Clay Lake (*black symbols*; bottom sediments grading into natural clay and silt). Sampling sites and corresponding symbols: outflow of chlor-alkali plant, ⊗; inflow of Wainwright reservoir, ○; outflow of Wainwright reservoir, △; Minnitaki, ●; Quibell, ▲; Clay Lake outflow, ◆. (From Jackson et al. 1982; republished with the permission of the Editor-in-Chief of Hydrobiologia)

source of pollution owing to dilution of toxic contaminants (Jackson et al. 1982). The results demonstrate vividly that in this system, as in others investigated by the writer, the methylation and bio-accumulation of Hg are controlled primarily by environmental factors affecting (1) the speciation and bio-availability of inorganic Hg and (2) the growth and activities of microbes capable of methylating the bio-available inorganic Hg species; in contrast, Hg methylation and bio-accumulation are usually independent of the total abundance of Hg in the environment within a wide range of total Hg concentrations and may even be *inversely* related to total Hg.

Deposition of fluvially transported clay and silt in riverine lakes also has an important bearing on the fate of Hg in the Wabigoon River system. Clay Lake, the first lake downstream from the source of pollution (Fig. 5.18), traps Hg-bearing detritus introduced into it by the river. The efficiency of heavy metal entrapment was found to increase with the standard *entropy* of metal ion hydration and is therefore higher for Hg than for Cu, Zn, Cd, or Fe (Jackson 1979).

Pronounced seasonal as well as spatial variations were seen in the Wabigoon River (Jackson et al. 1982). Particulate Hg in the river water was most abundant during the spring flood owing to bottom scour and re-suspension of fine sediment, but 98–100% of it was nonextractable by

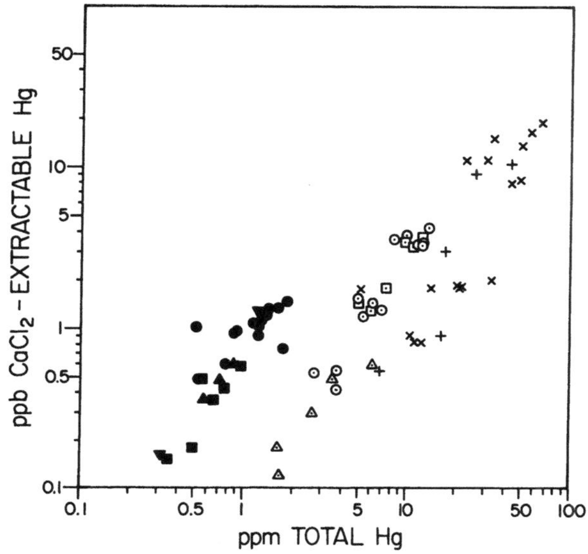

Fig. 5.21. Relationships between 0.5 M $CaCl_2$-extractable Hg and total Hg concentrations in bottom sediments from different regions of the Wabigoon River system (Fig. 5.18) in the summer of 1978. (1) Riverbed between chlor-alkali plant at Dryden and inflow of Wainwright Reservoir (wood-chip deposit, + ; underlying clay-silt sediment, ▲); (2) Wainwright Reservoir (wood-chip deposit, X ; underlying clay-silt sediment, ▼); (3) Clay Lake (east basin, ⊙; west basin, ⊡); (4) clay-silt sediment from riverbed downstream from Clay Lake (△); and (5) clay-silt sediment from Ball Lake (north basin, ■; south basin, ●). (From Jackson and Woychuk 1980b)

mild solvents. In the mid-summer, a higher percentage of the particulate Hg was extractable with solvents, as also observed elsewhere (Jackson 1988b), and the concentrations of both inorganic Hg and methyl Hg in the operationally defined "solution" phase of the river water were higher, as was the percentage of Hg in the form of methyl Hg (Fig. 5.19C). These observations probably reflect an increase in the solubility, bio-availability, and methylation rate of inorganic Hg from spring to summer owing to intensified microbial activity, as well as factors such as the low flow rates, minimal dilution by runoff, high rates of evaporation, and low concentrations of suspended particles which characterise the hydraulic regime in the summer. The affinities of Hg for particular constituents of the suspended matter (organic substances, Se, and Fe and Mn oxides) also varied seasonally *and*, in part, spatially as well, suggesting fluctuations in the partitioning of Hg between different binding agents; moreover, the binding agents apparently discriminated between inorganic Hg and methyl Hg (Jackson et al. 1982).

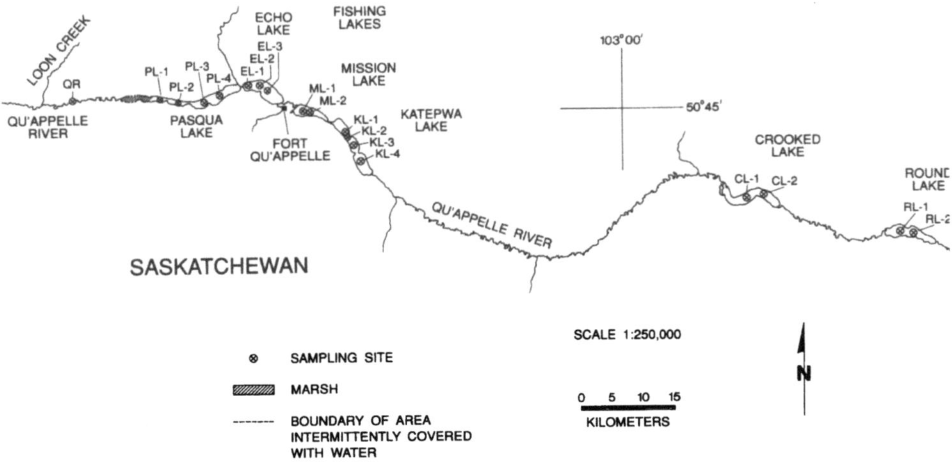

Fig. 5.22. Map of part of the Qu'Appelle River, Saskatchewan (Canada) and its chain of lakes showing location of sampling sites. Each lake site represents a major basin of deposition. The river flows from west to east, and the sources of Hg pollution (not shown) are located many kilometers upstream from Pasqua Lake, the first lake in the series. The Hg, which originated in two nearby cities, was discharged into streams flowing into the Qu'Appelle River, either directly or by way of a tributary (the Moose Jaw River; Jackson 1986). (From Jackson 1993b; republished with the permission of the editors of the Water Pollution Research Journal of Canada)

The Moose Jaw River–Qu'Appelle River system, including a chain of extremely eutrophic riverine lakes, in the prairie region of Saskatchewan (Fig. 5.22) provides another example of spatial and temporal variations in environmental conditions affecting the speciation, binding, release, transport, and deposition of Hg (Jackson 1986, 1993b, c). Comparison with the Wabigoon River system may be instructive, as the two systems differ radically in physicochemical and biological characteristics and yet display certain parallels. The particulate inorganic Hg content of the river water was highest during the high-flow episode of the spring flood owing to resuspension of contaminated clay-sized sediment; it then dropped sharply and was minimal during the slack-water period of summer and autumn. However, methyl Hg levels were minimal during the spring flood and then increased during the slack-water period in response to a rise in phytoplankton production, which stimulated the growth of methylating bacteria by furnishing organic nutrient substrates (Jackson 1986). Thus, total Hg and methyl Hg were inversely related. In these respects, the river system was comparable to the Wabigoon River system. The chief carrier of inorganic Hg in the river system appears to be Fe oxide, which probably forms coatings on clay and silt particles (Jackson 1986). On entering

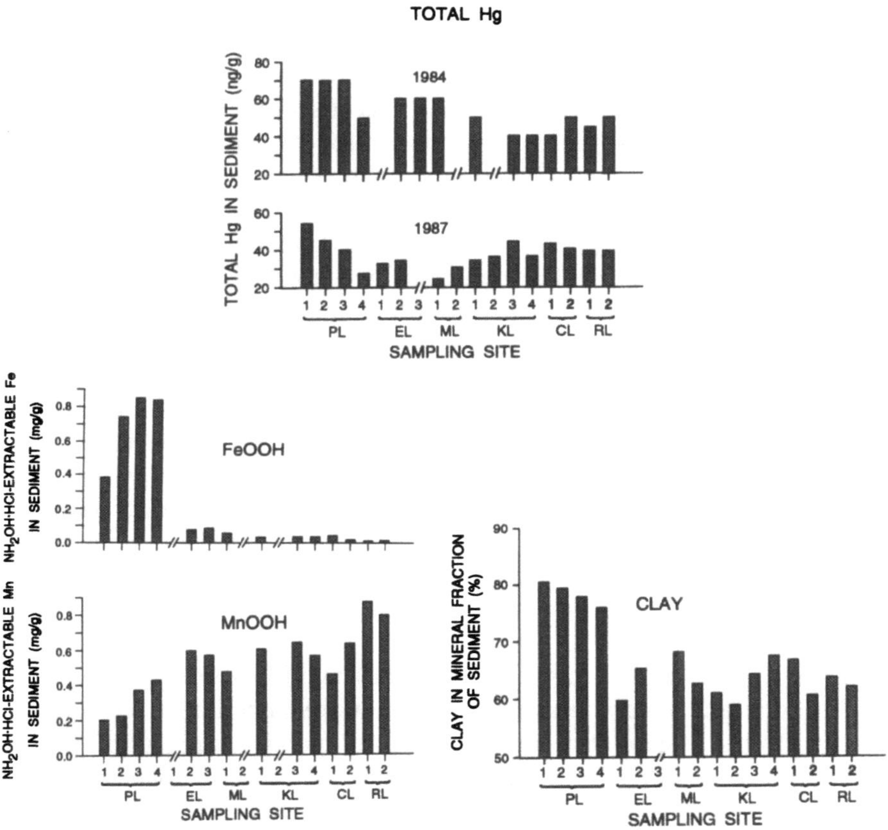

Fig. 5.23. Variations in the total Hg, $NH_2OH \cdot HCL$-extractable "FeOOH" and "MnOOH", and clay concentrations in bottom sediments of the main basins of the Qu'Appelle River lakes (Fig. 5.22). The data are arranged in the order of occurrence of the sampling sites, going in the downstream direction (from *left* to *right*), away from the source of Hg pollution. Years in which samples were collected and analysed: *total Hg*, 1984 and 1987; *FeOOH* and *MnOOH*, 1984; *clay*, 1987. The sampling sites are identified by abbreviations of the names of the lakes (e.g. *PL* for Pasqua Lake) and by *numbers* assigned to the lake basins (Fig. 5.22). (From Jackson 1993b; republished with the permission of the editors of the Water Pollution Research Journal of Canada)

Pasqua Lake, the first of the chain of lakes through which the Qu'Appelle River flows (Fig. 5.22), the particles settle out, contributing to the buildup of a large delta. Systematic variations in the composition of sediments in the main basins of the lakes from the upstream end to the downstream end of the system reveal that Pasqua Lake is a highly efficient trap for detrital Fe oxide and clay minerals but, surprisingly, is much less effective in trapping Hg (Jackson 1993b; Fig. 5.23). The efficiency of Hg, Cd, Cu,

Zn, and Fe entrapment by Pasqua Lake sediment was found to decrease with the standard *enthalpy* of metal ion hydration and was therefore lower for Hg than for any of the other metals except Cd (Jackson 1979). This is practically the opposite of what was observed in Clay Lake, where, as we have seen, the process was controlled by standard *entropy*; the contrast reflects the profound difference in physicochemical conditions between the two lakes (Jackson 1979). On being deposited in the extremely reducing bottom environment of the lake, the Fe oxide apparently undergoes at least partial reduction and solubilisation, releasing its bound Hg; the Hg is then partitioned among organic substances and sulfides, and much of it is gradually transported down the river system, probably on organic particles, and is dispersed through the other lakes in the series, whereas most of the Fe is retained in Pasqua Lake. In contrast to the extractable "Fe oxide" fraction, "Mn oxide" tends to increase progressively in the downstream direction throughout the system (Fig. 5.23), probably owing to the combined effect of dilution by detrital clay at the upper end of the system, greater solubility of Mn than Fe under the existing redox conditions, and accumulation of Mn oxide in the well aerated, environment of the downstream end (Jackson 1993b, c).

Changes in pH (see Sect. 5.4.1.4 to 5.4.1.6). Acidification of ill-buffered soils and inland waters causes desorption of metal cations from colloids, leading to the release of many sorbed heavy metals into solution (Förstner and Patchineelam 1976; Förstner 1980, 1991; Khalid 1980; Pickering 1980a, b; Brümmer 1986; Karlsson et al. 1987; Tessier et al. 1989), whilst neutralisation of acids fosters sorption of metals (Loganathan et al. 1977; Davis and Leckie 1978; Rickard and Nriagu 1978; Thornton 1979; Leckie et al. 1980; Pickering 1980a, b; Abd-Elfattah and Wada 1981; Barrow et al. 1981; Benjamin and Leckie 1981; Jarvis 1981; Kinniburgh and Jackson 1981; Brümmer et al. 1983; Bruemmer et al. 1986; Tessier et al. 1989; Duarte et al. 1991; Madrid et al. 1991; McBride 1991b; Schmitt and Sticher 1991; Swift and McLaren 1991; Manceau et al. 1992; Tessier et al. 1993). As metals differ in their tendencies to be solubilised by acids, acidification of colloid–metal complexes leads to selective leaching. For example, sorption of Pb and Cr by soil minerals was found to be less pH-sensitive than sorption of Cu, Cd, Zn, Ni, and Co (Andersson 1977). The role of pH in the biogeochemistry of Hg is in a special class because methylation and demethylation as well as the sorption and desorption of inorganic species are involved (see below). It must also be remembered that the effects of pH on anions are the opposite of their effects on cations. Thus, the sorption of anionic metal species by mineral colloids is increased by acidification but decreases with increasing pH (Hingston et al. 1968; Calder

1988). This applies to certain metals and metalloids that are usually in the form of anions, such as V, Mo, Cr (VI), As, and Se, and also to metals such as Hg, Pb, and Co which form anionic species under highly alkaline conditions but are otherwise in the form of cations or uncharged species.

The release of bound metals into solution by acidification depends not only on the properties of the metals but also on the nature of the binding agents. Metals are more readily leached from mineral colloids than from organic matter, which is therefore the most (or indeed the only) effective binding agent at low pH (Andersson 1977, 1979; Schuster 1991; Tessier et al. 1996; Fig. 5.10).

Increased mobilisation and leaching of many metals (e.g. Cd and Zn) is one of the effects of the widespread problem of "acid rain" in poorly buffered soils and lakes (Förstner 1980, 1991; Tessier et al. 1989). Acidification promotes the accumulation of methyl Hg by fish in poorly buffered lakes, suggesting increased mobilisation of Hg; but this process is poorly understood, and it could be a complex function of several phenomena, including effects of pH and other factors on the Hg methylating and demethylating activities of microbes (Jackson 1987; Richman et al. 1988; Winfrey and Rudd 1990).

A more localised but nonetheless important environmental problem due to acidification is acid mine drainage. Oxidation of sulfide minerals by reaction with dissolved O_2 with the aid of bacteria such as *Thiobacillus ferro-oxidans* results in production of H_2SO_4, causing extremely acidic conditions and the leaching of heavy metals (Weatherley et al. 1980). Consequently, waters draining tailings ponds and other waste disposal sites containing metal sulfides undergoing oxidative weathering have low pH values and are charged with dissolved metals (Förstner and Wittmann 1976; Förstner 1980; Karlsson et al. 1987). Neutralisation of the acid following discharge of the wastewaters into rivers or other natural waters leads to precipitation of Fe and Al oxides, which scavenge heavy metals by coprecipitation and sorption (Lee 1975; Karlsson et al. 1987), although the pollutants may be widely dispersed if the receiving waters are too poorly buffered to bring about a sufficient rise in the pH immediately (Förstner and Wittmann 1976). Karlsson et al. (1987) reported that during the discharge of tailings pond effluent into a stream the quantities of heavy metal taken up by the flocculated oxides as the pH increased varied in the order Pb > Cu > Cd > Zn. In a case investigated by Jackson (1978, 1979), acidic metal-polluted water draining the tailings pond of a base metal mine is discharged into a nearby lake, which is used as the receiving basin for municipal sewage effluents. In the surface water of the lake photosynthetic activities of algal blooms generated by nutrients from the sewage create alkaline conditions, neutralising the acidic mine effluent.

The high pH combined with a high concentration of mainly organic suspended particles promotes sorption of metals by the particles, hastening the transfer of the metals to the bottom, where they are bound strongly by sulfides generated by putrefaction of organic matter accompanied by SO_4^{2-} reduction.

Oxidation and Reduction (see Sect. 5.4.1 and 5.4.2). Major changes in metal mobility and the distribution of metals among environmental compartments occur at boundaries between O_2-rich and anoxic or O_2-poor domains, or between "oxidising" and "reducing" environments ($Eh > 0$ and $Eh < 0$, respectively). The most common interfaces or transition zones of this kind include (1) the metalimnion, or thermocline, of a productive stratified lake (or the equivalent boundary in certain restricted marine basins such as the Black Sea and deep fjords) in which the well-aerated epilimnion overlies an O_2-poor, anoxic, or highly reducing hypolimnion where organic matter generated by primary production in the epilimnion undergoes decomposition accompanied by consumption of O_2 and production of H_2S and other reducing agents; (2) the sediment–water interface in lakes in which anoxic sediments are overlain by oxygenated water (at least, during certain times of year); (3) the surface of any mass of labile organic matter (e.g. a plant fragment or the dead body of an animal) undergoing microbial decomposition within an oxygenated environment (e.g. soil or well aerated surficial sediment), thereby creating a temporary anoxic or reducing microenvironment; and (4) the surface of waterlogged soil. Some anoxic or reducing environments are perpetually isolated from the overlying oxygenated ones (as in the Black Sea and meromictic lakes), whereas in other cases there is periodic (often seasonal) oxygenation of anoxic or reducing environments (as with the spring and autumn overturn of lakes).

In the presence of free O_2, heavy metals in sediments and soil are mostly bound to Fe and Mn oxides and organic (especially humic) matter, and (less effectively) to clay minerals. Under reducing conditions oxides are reduced and dissolved, and heavy metals are mainly bound by sulfides and organic matter (including many nonhumic compounds that are much less labile in reducing than in oxidising environments). In the presence of H_2S, many heavy metals have a strong tendency to form highly insoluble sulfides (HgS, CuS, etc.), but since Fe is, as a rule, far more abundant than any heavy metal, they are usually coprecipitated or sorbed by amorphous FeS instead of forming discrete metal sulfide phases. Organic sulfide complexes and soluble metal sulfide species may also be formed. Even in highly reducing environments where H_2S abounds, metals seem to be partitioned between the inorganic sulfide

(FeS) phase and organic matter (the degree of preference for the sulfide phase depending on the properties of the metal) (Jackson 1978, 1979); generally speaking, even metals that form highly insoluble sulfides differ among themselves in their tendency to be immobilised in the presence of H_2S (Spencer et al. 1972; Jackson 1978, 1979; Jackson and Bistricki 1995). As Fe oxides, Mn oxides, FeS, and organic matter have different relative affinities for heavy metals (see Sec. 5.4.1.7), the partitioning of metals is strongly affected by variations in oxidation-reduction conditions, which regulate the proportions and absolute concentrations of these metal-binding substances in the environment (Jackson 1978, 1979; Jackson and Bistricki 1995).

When reduced sediments in a body of standing water are exposed to dissolved O_2 at the sediment-water interface, FeS and other sulfides, along with interstitial dissolved Fe(II) and Mn(II), are oxidised, and highly insoluble Fe and Mn oxides are precipitated. In lakes, therefore, the sediment-water interface is commonly marked by a thin surface layer of oxides formed by oxidation of upward-diffusing dissolved Fe(II) and Mn(II) as they come into contact with dissolved O_2 in the overlying water, and the same phenomenon occurs in waterlogged soil (Andersson 1979). Oxidation of sulfides releases heavy metals bound to them, but these metals, together with dissolved metal and metalloid species diffusing upward through the reduced zone, are immediately trapped and accumulated by the Fe and Mn oxides forming at the sediment surface and are thus at least partially prevented from diffusing into the water column (Lee 1975; Andersson 1979; Förstner 1980; Young et al. 1980; Salomons and Baccini 1986; Belzile and Tessier 1990; Morgan and Stumm 1991; Petersen et al. 1993; Zhuang et al. 1994). Similarly, vigorous aeration (e.g. strong current action over the bottom of a riverine lake or an area of sea floor whose sediments are composed mainly of mineral detritus rather than organic matter) promotes the precipitation of Mn-Fe oxide nodules on the sediment surface and Mn-Fe oxide encrustations on submerged rock surfaces and sediment particles. These oxide deposits are commonly enriched in sorbed and coprecipitated trace elements, although, for unknown reasons, certain lacustrine nodules are deficient in trace elements (Lee 1975; Leckie and Davis 1979; Boust and Joron 1986). As we have seen, the separate Mn and Fe oxide phases of these nodules and coatings have distinctly different trace element assemblages.

Comparable processes occur at the boundary between oxygenated and anoxic zones in the water column of a lake or restricted marine basin. In a stratified lake with anoxic or O_2-poor hypolimnion water, dissolved Mn(II) diffusing upward into the metalimnion from the hypolimnion is precipitated as Mn oxide on being oxidised by dissolved O_2 diffusing downward from the epilimnion (Lee 1975; Jackson 1993b); the same phenomenon occurs in the Black Sea and in fjords (Spencer et al. 1972; Berrang and Grill 1974). Similarly, Fe and Mn oxides are precipitated on surfaces of suspended particles, plankton, and aquatic insects, in natural waters (Fowler 1977; Dunlap et al. 1983; Dunlap and Walne 1985, 1987; Barnes et al. 1986; Hare et al. 1991; Cain et al. 1992; Hare 1992; Jackson and Bistricki 1995), and on bacterial cells (Urrutia and Beveridge 1993). These oxides remove dissolved trace elements from the water; but, as with Mn–Fe nodules, the process may be selective. Fe and Mn oxide deposits on exposed surfaces of plankton hard parts in several Canadian lakes polluted with heavy metals differ from each other in trace element composition, Fe oxide selectively accumulating Cu and As whilst Mn oxide preferentially binds Zn (Jackson and Bistricki 1995).

What happens when oxidised sediments are reduced? Trace elements bound to Fe and Mn oxides are released into solution during reduction and dissolution of the oxides (Wang et al. 1989; Belzile and Tessier 1990; Francis and Dodge 1990; Jackson 1993b), but they are likely to be immobilised again by sulfides and organic matter (Förstner 1980; Young et al. 1980). Nevertheless, small quantities of operationally defined "dissolved" metal (e.g. Hg) species liberated from reduced oxides have been detected in the pore water of reduced sediments and in overlying anoxic water (Gobeil and Cossa 1993; Jackson 1993b).

In brief, we have seen that in both the oxidation of reduced sediment and the reduction of oxidised sediment, bound heavy metals are released and then largely immobilised again by different solid phases which are stable under the new oxidation-reduction regime. A question, then, of great ecological importance arises: are heavy metals more soluble (and therefore possibly more bio-available) in an oxidising or a reducing environment? From the standpoint of metal binding and release, what are the *net* effects of oxidation and reduction? A number of independent investigations have shown that exposure of reduced Cd-contaminated marine or freshwater sediments to oxidation increases the solubility of the Cd owing to oxidation of sulfides, resulting in accelerated release of Cd into the overlying water and an increase in the proportion of the sediment-bound Cd that is weakly bonded and exchangeable (Förstner 1980, 1991; Zhuang et al. 1994). As a rule, oxidising conditions seem to promote the mobilisation of Cd regardless of whether the pH is acidic, neutral, or alkaline (Förstner 1980). The work of Zhuang et al. (1994) is particularly interesting, because it demonstrated that aeration of reduced sediment caused a net increase in the proportion of dissolved Cd in the environment despite concurrent precipitation of Fe and Mn oxides which scavenged part of the solubilised Cd. Similarly, other heavy metals are

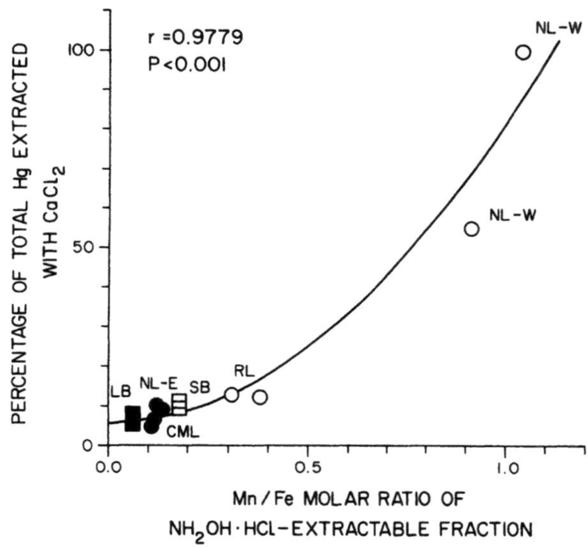

Fig. 5.24. Relationship between the percentage of Hg extractable with 0.5 M $CaCl_2$ and the Mn/Fe ratio of the $NH_2OH \cdot HCl/HNO_3$-extractable oxide fraction in sediments from different sites in hydroelectric reservoirs (southern Indian Lake and lakes of the Notigi Reservoir complex) in Manitoba, Canada. *Letters* associated with *symbols* represent names of sampling sites. *White symbols* Sites where the water is vigorously flushed and well oxygenated by fluvial currents; *black symbols* low-energy, less effectively aerated backwater sites. *Round symbols* Notigi Reservoir sites; *square symbols* southern Indian Lake sites. (From Jackson 1988b; republished with the permission of the Editor of the Canadian Journal of Fisheries and Aquatic Sciences)

mobilised by oxidation (Förstner 1991). Some workers have maintained that Zn is most soluble under reducing conditions because reduction liberates it from Fe and Mn oxides, but Young et al. (1980) drew the opposite inference, i.e. that although Zn and other metals are released into solution by reduction of oxides at low Eh levels, sulfides may render the metals much less soluble than they were under oxidising conditions. Furthermore, research on contrasting riverine lake and reservoir environments in Canada has shown consistently that inorganic Hg in sediments is more readily solubilised, and hence is more available for methylation and bio-accumulation, in well oxygenated environments than in anoxic or reducing ones, probably owing to oxidation of sulfides and selenides (Jackson 1988b, 1991, 1993c; Fig. 5.24). To summarise, it would seem that metals are usually more soluble in oxidising than in reducing environments.

In the case of Hg, however, the effects of dissolved O_2 and Eh are exceptionally complex because the microbial synthesis and decomposition of methyl Hg, which these parameters affect, has to be taken into account. The net rate of methyl Hg production is, in part, a compromise between two opposing effects of oxidation-reduction conditions. In well aerated environments, inorganic Hg availability is greatest but the methylating activities of microbes are relatively low, whereas in anoxic or reducing environments inorganic Hg is less available but methylating activity is maximised (Jackson 1988b, 1993c). The result is a complex trade-off. The net rate of methyl Hg production is likely to be highest in anoxic and reducing environments where there is an ample supply of organic nutrient substrates, because intense activity of methylating microbes compensates for the relatively low availability of inorganic Hg (Jackson 1987, 1988b, 1993c); but in well aerated environments the net rate of methyl Hg production may be anomalously high owing to exceptionally high inorganic Hg solubility, even if methylating activity and total Hg concentrations are both low (Jackson 1993c). In any aquatic environment, methyl Hg production is a function of both the supply of available inorganic Hg and the activities of methylating and demethylating microbes; it is, therefore, the net result of many physicochemical and biological variables simultaneously reinforcing or cancelling each other's effects.

5.5 Participation of Minerals in Trace Element Speciation

Besides sorbing trace elements, oxides and clay minerals may alter them by taking part in their speciation reactions. There may be a direct interaction between the element and the mineral, as in the oxidation of trace element species by oxides or a catalytic reaction on the surface of a clay crystal; or the interaction may be indirect, as in the effects of minerals on bacteria that mediate speciation reactions. If we were to pursue this latter topic to the fullest extent, there would be no end to it, as biological activities control the geochemical processes of the biosphere, metal speciation included, and are themselves profoundly affected by the products of their activities, including colloidal minerals. For the purposes of this review we will just touch briefly on two subjects: oxidation of trace element species by oxides and effects of colloidal minerals on microbes that methylate and demethylate Hg.

5.5.1 Reactions of Trace Elements with Fe and Mn Oxides

Laboratory experiments have proven that Fe and Mn oxides may oxidise certain trace element species abiotically. The importance of such reactions in natural soil and aquatic environments, where chemical transformations are generally controlled by micro-organisms, is unknown; but since they involve the conversion of highly toxic species to relatively non-toxic ones or vice versa, they could be of ecological importance if, in fact, they occur in nature.

Oscarson et al. (1981 a–c, 1983) and Moore et al. (1990) reported oxidation of As (III) to As (V) by Mn oxide. De Vitre et al. (1991) also found that Fe oxide may accomplish this transformation. Since As (III) is much more toxic than As (V), the oxidation of As (III) to As (V) by oxide minerals could have a significant detoxifying effect on As. Yet the oxidation of certain trace elements may be undesirable. Thus, Mn oxides may oxidise Cr (III) to Cr (VI), thereby greatly increasing the toxicity of the Cr (Fendorf and Sparks 1991). Hem (1978) has postulated that Mn oxides may oxidise other metals too, including Pb, Ni, and Co. Jackson (1989) demonstrated spontaneous abiotic destruction of methyl Hg by synthetic and natural Mn oxides (but not Fe oxides). In the presence of lake sediments with viable microbial communities, however, Mn oxides either inhibited or enhanced the breakdown and formation of methyl Hg, depending on experimental conditions (Jackson 1989). Therefore, it is doubtful whether abiotic demethylation of Hg by Mn oxides plays a significant part in Hg speciation in natural environments, where microbes hold sway.

Heavy metals can alter oxides as well as being altered by them. Thus, Cd (II) was found to increase the rate of Mn (IV) oxide formation from Mn (II) (Hem and Lind 1991). But sorption of Cu by Mn (III) and Mn (IV) oxide particles whose surfaces were undergoing reduction to Mn (II) resulted in displacement of Mn (II), enhancing the release of Mn into solution (Traina and Doner 1985).

5.5.2 Effects of Minerals on Hg Methylation and Demethylation

Clay- and silt-sized mineral detritus transported into lakes by fluvial action or the erosion of shoreline material inhibits the activities of Hg-methylating microbes in surficial sediments, thereby limiting the accumulation of Hg by fish (Jackson 1987, 1988 b, 1991, 1993 b, c). Probable reasons for this include (1) reduced light penetration owing to turbidity, limiting the primary production of labile organic nutrient substrates utilised by the microbes; (2) rapid burial and dilution of organic matter and other nutrients; and (3) the smothering of microbes by prevention of the exchange of dissolved nutrients and wastes between the sediment and water (Jackson 1988b, 1993b).

However, the results of laboratory experiments on the effects of clay and oxides on the Hg methylating and demethylating capabilities of indigenous microbial communities in sediment samples under different defined conditions suggest that the effects of colloidal minerals on the microbes are more complex and more specific than might be supposed from the field data (Jackson 1987, 1989, 1995). In these experiments, replicate sediment–water mixtures with and without added colloids ("experimental" and "control" systems, respectively) were incubated in an atmosphere of N_2 under different conditions (e.g. with and without nutrient enrichment or buffering), and the results were compared. Depending on the nature, abundance, and surface chemistry of the mineral colloid, the experimental conditions, and the source of the sediment (and therefore the species composition of the microflora), the mineral particles either inhibited or promoted the Hg transformations or had little or no effect, or they exerted a succession of different effects as the experimental conditions (e.g. the abundance of the added colloid) varied or as the incubation time increased. Most of the results suggested selective effects on particular kinds of microbes rather than general inhibition or stimulation of microbes or effects due to the binding of Hg by the minerals. A major reason for the complexity and variability of the effects of the minerals is that microbial communities comprise many species and strains, including different varieties of Hg methylating and demethylating microbes, that differ widely from one another in their activities, ecological requirements and limits of tolerance, and interact with each other in various complex ways ranging from symbiosis and mutualism to antagonism and competition. The minerals probably altered the species composition of the active portion of the microflora, affecting the course of ecological succession and thereby changing the dynamic balance between methylation and demethylation. As a rule, clay and silt interfered with methyl Hg production, but their effects differed qualitatively from one lake to another, probably reflecting differences in the nature of the microbial populations. In experiments on sediment samples from a Boreal forest lake, individual clay minerals and varved silty clay (Pleistocene glacial lake sediment of heterogeneous mineralogy from a shoreline deposit) had no effect, or a marginal inhibitory effect, on methylation but strongly promoted subsequent demethylation (when demethylators supplanted methylators as the dominant Hg-transforming microbes in the community; Fig. 5.25A). In sediment from an extremely eutrophic prairie

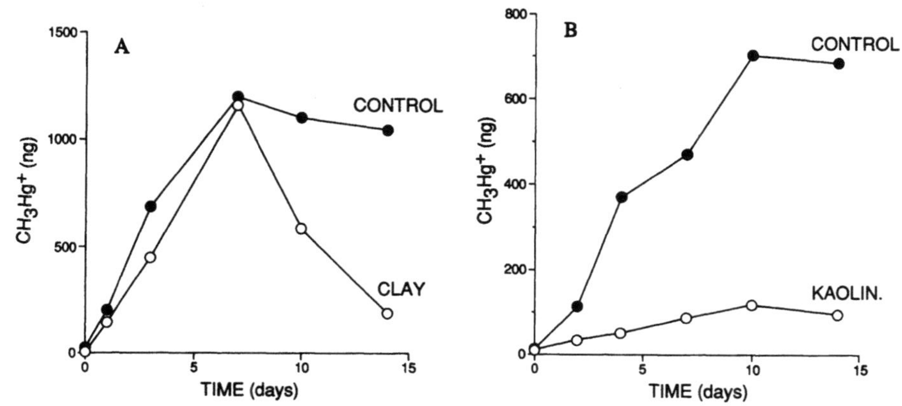

Fig. 5.25 A, B. Results of experiments on the effects of clay on microbial production and decomposition of methyl Hg (CH_3Hg^+) in lake sediments. The net quantity of methyl Hg produced during incubation under standardised conditions is plotted against incubation time; the data represent **A** sediment from East Mynarski Lake (Boreal forest zone, Manitoba, Canada) with and without added Pleistocene silt-clay sediment (*clay*), and **B** Pasqua Lake (prairie zone, Saskatchewan, Canada; Fig. 5.22) with and without added kaolinite (*kaolin.*). *Controls* were systems without added clay. All systems were amended with $CaCO_3$ (for buffering), pulverised plant particles (for nutrient enrichment), and dissolved $HgCl_2$, and they were incubated under N_2. Experimental systems with added clay, ○; control systems, ●. (From Jackson 1987, 1989; republished with the permission of the Longman Group Ltd. and the Editor of Applied Organometallic Chemistry)

lake, however, clay minerals strongly inhibited methylation but did not stimulate subsequent demethylation (Fig. 5.25 B). Thus, in some lakes the suppression of methyl Hg production by clay could be due to selective stimulation of demethylators, but in other lakes it could be due to inhibition of methylators.

The role of coatings on clay particles was examined by performing experiments using mineral samples with and without their natural coatings, and also with and without artifical oxide coatings; in addition, the quantities of natural Fe and Mn oxide on the clay surfaces were determined with the aid of solvent extraction and were correlated with methylating activity. The results demonstrated that oxide coatings played critically important ecological and biogeochemical roles. Fe oxide often promoted methylation, and Fe oxide coatings on clay tended to abate the inhibitory effect of the clay on the methylating microbes. Compared with montmorillonite in its native state, montmorillonite whose oxide coatings had been removed by citrate/dithionite extraction depressed methylation (Fig. 5.26 A), and removal of oxide coatings from the above-mentioned

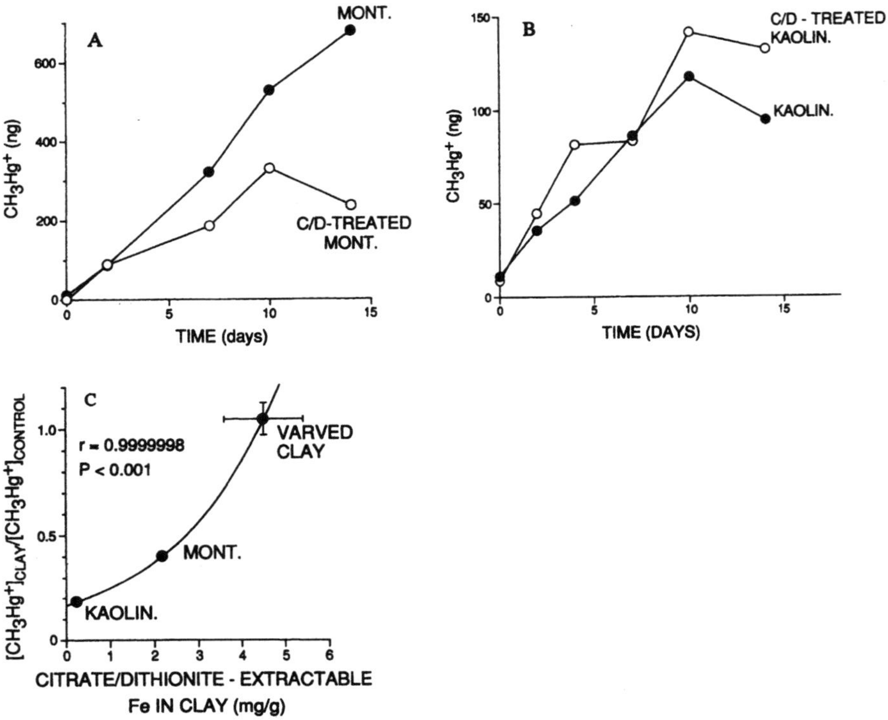

Fig. 5.26 A–C. Experimental results demonstrating that Fe oxide coatings on clay strongly affect microbial production of methyl Hg (CH_3Hg^+) in lake sediments. Experimental conditions, materials, and design are described in the caption for Fig. 5.25. The data show **A** methyl Hg production by Pasqua Lake sediment in the presence of montmorillonite (*mont.*) (●) and in the presence of citrate/dithionite (*C/D*)-extracted (i.e. oxide-free) montmorillonite (○); **B** methyl Hg production by Pasqua Lake sediment in the presence of kaolinite (*kaolin.*) (●) and C/D-extracted kaolinite (○); and **C** the quantity of methyl Hg produced by clay-amended Pasqua Lake and East Mynarski Lake sediment relative to sediment without added clay during 7 days of incubation plotted against the abundance of citrate/dithionite-extractable Fe oxide on the clay particles. *Varved clay* in **C** is the same as "clay" in Fig. 25A. (From Jackson 1989; republished with the permission of the Longman Group Ltd. and the Editor of Applied Organometallic Chemistry)

varved silty clay depressed both methylation and demethylation (Jackson 1989). In contrast, the stripping of oxides from kaolinite (the clay with the lowest oxide content) enhanced methylation (Fig. 5.26 B). Experiments measuring the inhibition of Hg methylation by kaolinite, montmorillonite, and varved silty clay showed that the observed effects of the mineral particles on methyl Hg production were entirely attributable to natural Fe oxide coatings on the particles and had nothing to do with the nature of the silicate minerals on which the coatings had been deposited (Fig. 5.26C). Some stimulation of methylating activity by Mn oxide as well as Fe oxide coatings on clay was observed, but larger amounts of Mn oxide (unlike Fe oxide) strongly suppressed methylation. On addition of organic nutrients, oxide coatings fostered methylation and impeded demethylation; but without nutrient enrichment the reverse tended to occur. In short, the effects of oxides and clay minerals on microbial methylation and demethylation of Hg are vitally important but are complex, variable, poorly understood, and not altogether predictable.

A further cautionary note about impurities in clay would be worthwhile. The natural silty clay used in some of the experiments contained calcite and dolomite as well as oxide coatings, and pH buffering by these carbonate minerals created mildly alkaline conditions which stimulated the activities of the Hg-methylating microbes. The beneficial effect of the carbonates on the microbes overcompensated for the mildly negative effect of the non-carbonate mineral particles, causing a net rise in the rate of methyl Hg production (Jackson 1987). To obtain valid measurements of the impact of the non-carbonate minerals, it was necessary to eliminate the effect of the carbonates by buffering all experimental and control systems with $CaCO_3$. In an independent parallel study, Hecky et al. (1987, 1991) performed field experiments to test the effect of the same silty clay on methyl Hg production in a major Canadian lake in which Hg has become a problem owing to recent impoundment, and into which large amounts of this same material are being eroded from a shoreline deposit. But they neglected to take pH buffering by the carbonate minerals into account in their experimental design and in their conclusions (even though it was manifested by their own raw data). As in the preliminary experiments of Jackson (1987), their experiments showed that the overall effect of the silty clay was to promote methylation, but they did not realise that this result was attributable solely to buffering by the carbonates associated with the clay. Thus, they drew invalid inferences and used their results as a basis for unfounded speculation about the impact of clay in nature. All the natural surface waters of the field area tend to have slightly alkaline pH values owing to widespread occurrence of calcareous glacial deposits, and so it is the negative impact of the silicates and other non-carbonate components, not the positive effect of the carbonates, that is seen in environments where there is heavy loading of eroded silty clay detritus (Jackson 1988b, 1991). Therefore, Jackson's $CaCO_3$-amended model sediment ecosystems with and without silty clay are reasonable approximations of the field environments.

A general review of the ecological impact of clay minerals, oxides, and humic matter on microbes has been published elsewhere (Jackson 1995).

5.6 Biological Consequences of Mineral-Trace Element Interactions

In aquatic and terrestrial ecosystems, colloids function as both sinks and secondary sources of trace elements. The binding of these elements by colloidal particles commonly limits or prevents their uptake by organisms; but release of the elements into solution permits biological uptake, if the solubilised species is bio-available and is not scavenged by some other binding agent (dissolved or particulate) that lowers its bio-availability. A change in environment (e.g. a drop in pH) may result in a sudden release of bio-available metals from binding sites (Pickering 1980b). Thus, colloids do not merely inhibit biological uptake of metals; they also facilitate uptake by providing reservoirs of bio-available metal species in the environment, and if the particles are ingested by an animal (e.g. a filter-feeding or deposit-feeding invertebrate), they serve as vehicles for direct conveyance of metals into the animal's body.

The binding and release of metals and metalloids by colloids have profound ecological significance, because these processes control the biological uptake, and hence both the nutritional and the toxic effects, of these elements, to the benefit or detriment of the organisms affected. Binding and release of inorganic species of certain elements (e.g. Hg) also regulates the availability of these species for methylation by microbes. Furthermore, chemical alteration of a potentially toxic trace element by reaction with a colloid may either increase or decrease the bio-availability and toxicity of the element. These phenomena depend on the nature and concentrations of the trace elements and colloids, the specific characteristics and ecological relations of the organisms, and a wide range of environmental factors (e.g. pH, Eh, and dissolved complexing agents).

5.6.1 Bio-availability of Metals as Related to Binding and Release

5.6.1.1 *The Bio-availability of Nutrient Trace Metals*

Clay minerals, oxides, and humic matter in soil are vitally important to land plants and other soil organisms for a number of reasons, one of which is that they bind essential trace metals reversibly, accumulating pools of exchangeable, bio-available metal species in the higher soil horizons and preventing them from being washed away or leached into the subsurface. H^+ ions and soluble chelating agents secreted by the rhizosphere microbes associated with plant roots release bound metal cations into solution, enabling the roots to absorb them (Jackson and Voigt 1971). Certain sorbed metal fractions are not readily available for biological uptake, however, because they are too strongly bonded to the colloids or because they are trapped inside colloidal particles. The result may be a trace element deficiency. The same principles probably apply, with modifications, in aquatic environments.

The bio-availability of a colloid-bound trace element varies with the nature of the colloid and the external conditions. The following examples of empirical observations will suffice to illustrate the great variety of possible effects: (1) Cu in soils composed chiefly of minerals is more available to plants than Cu in peaty soils (Thornton 1979); (2) Zn sorbed to kaolinite or illite was found to be more available to plants than Zn sorbed to montmorillonite (Pickering 1980b); (3) the availability of clay-bound Zn to tomato seedlings was found to decrease in the order kaolinite > montmorillonite > illite > vermiculite, and added $CaCO_3$ increased the availability (Sabet et al. 1975) even though alkalinity is conducive to sorption, not desorption, of Zn (suggesting to the present writer that pH buffering by the $CaCO_3$ stimulated microbial activities, which intensified the desorption process); (4) the availability of Mo to plants may be controlled by the degree of crystallinity of Fe oxide and by pH (Karimian and Cox 1979); (5) sorption of Cu and other heavy metals by Mn oxide precipitates whose surfaces are undergoing reduction may enhance the release of Mn into solution, thereby promoting uptake of Mn by plants (Traina and Doner 1985); and (6) ingestion and solubilisation of suspended Fe oxide particles by marine protozoans renders the Fe more available for uptake by phytoplankton, thus ameliorating Fe deficiency (Barbeau et al. 1996).

As pointed out by Swift and McLaren (1991), there have been few studies of the relations between trace metal sorption by soil and uptake by plants, and the existing data are not altogether helpful. Many research results reveal inconsistencies, such as a correlation between the sorption and bio-accumulation of Zn but no such correlation for Cu (Swift and McLaren 1991). Given the wide range of factors which have a bearing on the bio-availability of trace elements, such inconsistencies are not surprising. In general, estimates of sorption parameters for soil have been unsuccessful in predicting the uptake of metals by plants growing in the soil, possibly because desorption is more relevant than sorption (Pickering 1980a; Swift and McLaren 1991). The idea that desorption kinetics are of primary importance in controlling biological uptake gains credence from empirical data indicating that the extraction of soil with aqueous solutions of mild reagents such as $MgCl_2$, $CaCl_2$, $Ca(NO_3)_2$, and the chelating agent DTPA, solubilises the soil-bound fractions of Cu, Zn, and certain other metals which are available to plants (Tiller et al. 1972; Latterell et al. 1978; Lindsay and Norvell 1978). As would be expected, these results prove that the more weakly bound metal fractions are those which are available for uptake by plant roots. In general, these operationally defined bio-available species, which make up only a fraction of the total metal content, consist of dissolved ions and organic complexes in the soil solution and ions sorbed to the exchange sites of fine particulate matter, notably clay minerals, silt, and organic matter (Pickering 1979, 1980 a, b; Berrow and Burridge 1991). But not all attempts to equate readily-extracted metal fractions with fractions available to plants have been successful (Jarvis 1981), reminding us that bio-availability depends on a complex assortment of factors and is not fully understood.

5.6.1.2
The Bio-Availability of Toxic Heavy Metals

Control of biological uptake and toxicity of metals by colloids. The literature abounds with examples of the effectiveness of mineral colloids in abating the toxicity of various heavy metals (e.g. Cd, Pb, Hg, Cu, and Ni) to organisms by removing the metals from solution. A series of experiments by the pioneer microbial ecologist Stotzky and his associates demonstrated the ability of different clay minerals to protect soil bacteria and fungi from toxic effects of Ni, Cd, Zn, and Pb (Babich and Stotzky 1977 a, b, 1979, 1982, 1983 a, b; Bewley and Stotzky 1983 a, b; Debosz et al. 1985; Stotzky 1986; Collins and Stotzky 1989; Fig. 5.27). Similarly, montmorillonite protected microbes in lake water from Ni poisoning (Collins and Stotzky 1989). Moreover, amorphous Fe, Mn, and Al oxides interfered with the uptake of metals by microbes, and addition of $FeCl_3$ to a *Chlorella* culture provided protection against Cu, probably owing to precipitation of Fe oxide (Babich and Stotzky 1983 a). Other workers, too, have shown that clay minerals and soil clay fractions are able to protect microbes and microbial enzymes (e.g. dehydrogenase) against inhibition by Pb, Hg(II), and Cd (Kidby et al. 1977; Olson and Panigrahi 1991; Wilke 1991), and freshwater sediment protected anaerobic bacteria against toxic effects of Hg(II) (Hamdy and Wheeler 1978). Similarly, Fe oxide and organic matter in sediments have been found to hinder the accumulation of Cd by lacustrine bivalves (Tessier et al. 1993), and clay and organic matter limit the uptake of metals from soil by plants (van Driel and Nijssen 1988).

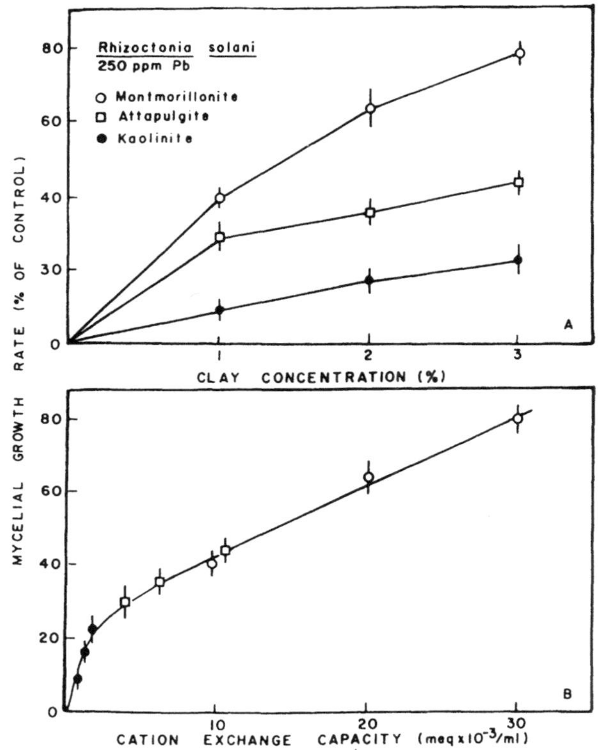

Fig. 5.27 A, B. Data showing the relative effectiveness of montmorillonite, attapulgite, and kaolinite in abating the toxicity of Pb to the fungus *Rhizoctonia solani*. Mycelial growth in the presence of an inhibitory ambient concentration of Pb has been plotted against **A** clay concentration and **B** the cation exchange capacity of the clay. (From Babich and Stotzky 1983a; republished with the permission of John Wiley & Sons, Inc. and H. Babich)

Effects of different colloids. In experiments comparing the detoxifying capabilities of different clay minerals, Stotzky and his coworkers found that montmorillonite invariably protected microbes more effectively than kaolinite did, and they attributed the superior detoxifying ability of montmorillonite to its high cation exchange capacity (Fig. 5.27). Their interpretation is consistent with their data (Fig. 5.27) and receives support from Wilke (1991), who discovered that inhibition of microbial dehydrogenase by Pb in soil declined as the cation exchange capacity of the soil increased, and from Haghiri (1974), who showed that uptake of Cd from soil by oat plants decreased as the cation exchange capacity of the soil increased. It seems doubtful, however, whether the inverse correlation between cation exchange capacity and bio-availability in any of these studies is a cause-and-effect relationship, as bio-availability was probably depressed primarily by specific sorption (see Sect. 5.4.1). The observation that montmorillonite had a greater detoxifying effect than kaolinite could be due to a larger number of specific sorption sites on montmorillonite, and on its oxide coatings, owing to the mineral's finer particle size and larger specific surface. Similar reasoning may be applicable to the work of Wilke and Haghiri as well. Besides, the exchangeable fractions of metals (e.g. Cd and Zn) in soil are the very fractions that are most readily available for biological uptake (Lorenz 1979; Davis-Carter and Shuman 1993).

In any event, colloids vary widely in their efficacy as detoxifying agents, as might be supposed from the differences in their sorption capacities and the binding energies of their sorption sites (see Sect. 5.4.1.7). Tessier et al. (1984) reported that amorphous Fe oxide was more effective than organic matter in preventing uptake of Pb, Cu, and Zn by freshwater pelecypods in freshwater environments polluted with mining and smelting wastes, and Gunn et al. (1989) observed, understandably, that Pb, Cu, Cd, and Zn were much more available for uptake by tubificid worms if they were sorbed to kaolinite than if they were coprecipitated with Fe oxide or bound to sewage sludge. Comparable to the findings of Gunn and associates are the data of Jackson and Woychuk (1980a, b, 1981) and Jackson et al. (1982) showing that inorganic Hg in the sediments of the Hg-polluted Wabigoon River system became more weakly sorbed, more readily desorbed by Cl^- ions or DTPA, and more available for methylation as the sediments graded from decomposing wood-chip debris into natural clay-silt mud with increasing distance downstream from the source of pollution (see Sect. 5.4.4.3). Paradoxically, however, the availability of Cd for uptake by oat plants was found to be greater in peaty soil than in clay-rich soil even though organic soils bind metals more strongly than clay soils (Pickering 1980a), demonstrating that a colloidal material's efficiency as a detoxifying agent is not necessarily a simple function of its metal-binding capability. Finally, it is worth mentioning that there is indirect evidence for selective effects of separate but coexisting Fe and Mn oxides in lacustrine environments. In some Boreal forest lakes Fe oxides appear to be particularly effective in preventing the uptake of inorganic Hg by certain kinds of benthic invertebrates (Fig. 5.28A), whilst Mn oxides are more effective for others (Fig. 5.28B), suggesting that particular oxide minerals and particular species of benthic animals take up certain Hg species in preference to others (Jackson 1988a).

Results of experiments by Stotzky and others demonstrating the detoxifying function of clay minerals have been confirmed by experiments with soil in its native state. Thus, Doelman (1978) reported that Pb suppressed microbial respiration (O_2 consumption) by 50% in sandy soil but only 15% in clay-rich soil and not at all in peaty soil. Using a different experimental design, Bewley and Stotzky (1983b) observed that fungi isolated from Cd-treated soil amended with kaolinite and montmorillonite were *more* sensitive to Cd than fungi from Cd-treated soil without added clay. This implies an effect of clay on natural selection in the

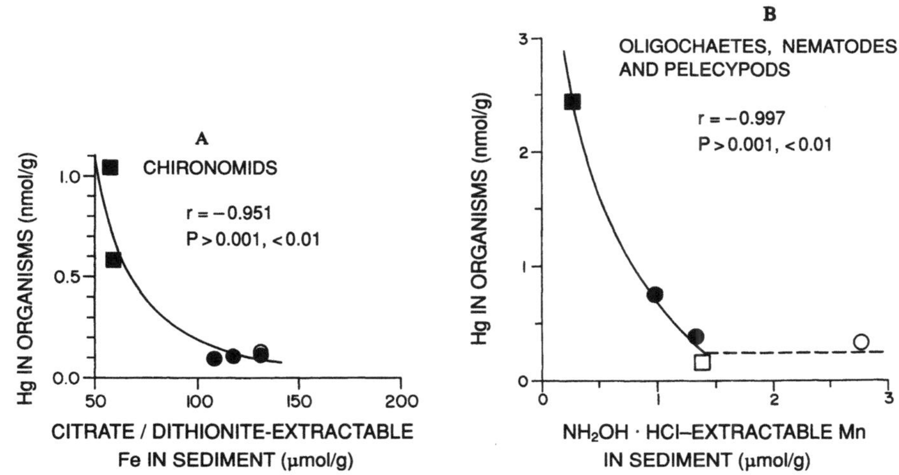

Fig. 5.28 A, B. Relationships between Hg concentrations in different kinds of benthic invertebrates and extractable Fe and Mn oxide concentrations in associated sediments in hydroelectric reservoirs of Manitoba, Canada. **A** Hg in chironomid larvae plotted against citrate/dithionite-extractable Fe; **B** Hg in oligochaetes, nematodes, and pelecypods (weighted average for all three groups combined) plotted against $NH_2OH \cdot HCl/HNO_3$-extractable Mn. Explanation of *symbols*: as in Fig. 5.24. (From Jackson 1988a: republished with the permission of the Editor of the Canadian Journal of Fisheries and Aquatic Sciences)

microbial community of the soil: only in relatively clay-poor soil did metal-tolerant microbial strains evolve, because the toxic effects of the metal were more severe in that environment.

As demonstrated by the experiments of Jackson (1989, 1995), colloids have selective effects on particular kinds of organisms, and their effects on organisms vary with environmental conditions. Experiments conducted by Bewley and Stotzky (1983b) showed that both montmorillonite and kaolinite protected soil fungi from a mixture of Cd and Zn, whereas only montmorillonite protected actinomycetes, and neither clay mineral protected eubacteria. Specific effects are seen in animals too: in several lakes polluted with heavy metals from a smelter, sedimentary Fe oxide depresses the accumulation of Cd in the liver of the white sucker (*Catostomus commersoni*; Fig. 5.29) but has no such effect on northern pike (*Esox lucius*) (Jackson et al. 1993); the difference may reflect the radical difference in the feeding habits of the two fish species (sucker being a bottom feeder whereas pike is a fish-eating predator). Lewis et al. (1972) reported variable, inconsistent results of experiments designed to test the ability of different clay minerals to protect a marine copepod from Cu. Although

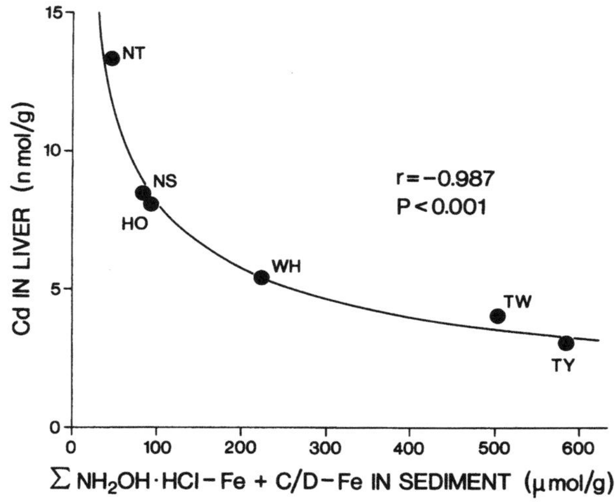

Fig. 5.29. Relationship between the Cd content of liver tissue in white sucker (*Catostomus commersoni*) (a bottom-feeding fish) and the sum of the $NH_2OH \cdot HCl$- and citrate/dithionite (C/D)-extractable Fe fractions (total FeOOH content) of associated sediments in lakes polluted by a base metal smelter at Flin Flon, Manitoba, Canada. The *letters* associated with the *symbols* represent names of lakes. (From Jackson et al. 1993; republished with the permission of Pergamon Press Ltd. (Elsevier)

kaolinite ameliorated Cu toxicity, the mineral itself had harmful effects on the organisms in two out of three tests at low Cu levels or with no Cu added. In addition, montmorillonite had either beneficial or harmful effects on the organisms depending on experimental conditions, whilst illite had no effect at all. The authors suspected that desorption of an unknown toxic impurity from montmorillonite accounted for the unfavourable effect of this mineral under certain conditions. These complex, contradictory results highlight the importance of taking surface impurities and environmental conditions into account in experiments involving clay minerals and other colloids.

Effects of Environmental Variables. The ambient pH has an important influence on the detoxifying role of clay. As neutral to mildly alkaline pH values promote the sorption of metal cations, metals such as Cd and Zn in soil may become less exchangeable, and hence less available for uptake by plants and less toxic to microbes, when the soil is limed or, at any rate, has a neutral or alkaline pH (Pickering 1980a; Babich and Stotzky 1982; Stotzky 1986; van Driel and Nijssen 1988; Davis-Carter and Shuman 1993). On the other hand, acidification of unbuffered soil and inland waters, a widespread phenomenon owing to acid rain, releases sorbed heavy metals from colloids, increasing the bio-availability, bio-accumulation rates, and toxicity of the metals (Förstner 1991; Tessier et al. 1993). (As explained above, acidification increases the bio-availability and potential toxicity of Hg but for different and more complex reasons involving microbial production of methyl Hg.) The desorption of metals from colloids on being transported into the marine environment by rivers (see Sect. 5.4.4.3) also increases the bio-availability and toxicity of metals (Förstner and Müller 1973). Cation competition is a factor too; for instance, Pb and Cu compete with Cd for sorption sites in soil, making Cd more available to plants; therefore, increasing the Pb content of soil may increase the Cd content of plants (Schmitt and Sticher 1991). Nevertheless, the interpretation of apparent interactions between different metals requires caution. Haghiri (1974) reported that addition of Zn to soil containing Cd increased the Cd content of plants growing in the soil, but this evidently resulted from suppression of plant growth, not from interaction between Cd and Zn. In general, an environmental change that increases the concentration of bio-available metal species in solution by causing desorption, or by interfering with sorption, may increase the toxicity of the metals, posing a threat to aquatic organisms and, if drinking water is affected, to humans (Förstner and Patchineelam 1976).

Methods for Estimating Bio-available Metal Fractions in Soil and Sediment. The more bio-available, and hence toxic, heavy metal fractions (not counting methyl Hg) in soil and sediment comprise the more loosely bound (e.g. exchangeable) species. As with bio-available nutritional trace elements, it is possible to estimate the bio-available fractions of toxic metals by means of certain mild solvent extractions. (Methyl Hg, despite its bio-availability, requires more rigorous extraction procedures.) Street et al. (1977) used DTPA to extract the Cd fraction available to plants growing in soil amended with Cd and sewage sludge; and, on isolating soil-bound Zn fractions by sequential extractions, Davis-Carter and Shuman (1993) found that exchangeable Zn correlated strongly with the Zn content of peanut plants growing in the soil. Methods developed for extraction of bio-available forms of metals from soil have been applied successfully to sediments, although a particular extractant may be suitable for some metals but not for others. Thus, inorganic Hg species available for microbial methylation can be extracted from at least some sediments with $CaCl_2$, but DTPA and dilute acetic acid have given unsatisfactory results for the same sediments, indicating that the Cl^- ion is required (probably because it forms stable soluble Hg complexes) (Jackson 1988b, 1991). Yet DTPA has been a useful reagent for extraction of sediment-bound Cd fractions which are toxic to sedimentary microbes and are available for accumulation in the liver of northern pike (Jackson et al. 1993). Moreover, sedimentary Pb, Cd, and Cu fractions available for uptake by tubificid worms may be extracted with NH_4 acetate/$CaCl_2$, a reagent used to isolate exchangeable fractions (Gunn et al. 1989). Since the availability of inorganic Hg in sediments is often limited chiefly by sulfides and selenides, the Hg/sulfide or Hg/Se ratio can be used as a semi-quantitative index of the availability of inorganic Hg for methylation (Jackson 1991, 1993c).

Biological Uptake of Sorbed Metals from Particles. Colloidal particles do not simply prevent organisms from taking up heavy metals from their environment: they may also serve as sources of metals for biological uptake.

Some organisms extract metals from colloids by altering their chemical environment, causing solubilisation of the metals. As with nutrient trace elements, the rhizosphere microbes of plant roots undoubtedly release toxic metal cations from nearby soil colloids by secreting H^+ ions and soluble chelating agents, whereupon the roots absorb the metals from the soil solution. Other microbes in both soils and sediments probably solubilise metals in essentially the same way. Laube et al. (1979) reported that freshwater cyanobacteria and green algae readily took up Cd and Cu sorbed to clay-sized fluvial sediment, accumulating considerable quantities of the metals, even though hardly any of the bound Cd or Cu dissolved in the water.

The other pathway for biological uptake of colloid-bound toxic metals, including metals bound to oxide coatings on plankton (Jackson and Bistricki 1995) and aquatic insects (Hare et al. 1991), is ingestion of the colloids by filter-feeding and deposit-feeding invertebrates, by bottom-feeding fish, and by predators that consume organisms with oxide coatings, followed by release of the metals through decomposition or leaching of the colloids during digestion. Molluscs, for instance, may take up Cd, Pb, and Cu from ingested particles (Lorenz 1979; Al-Sabri et al. 1993). Al-Sabri et al. (1993) found that filter-feeding freshwater snails took up toxic quantities of Cu from clay minerals, both by swallowing the clay and by transporting clay particles by means of ciliary currents toward the floor of the mantle cavity, where they were engulfed by epidermal cells or stripped of their sorbed Cu by H^+ ions in an acidic microenvironment created at the animal's body surface. Cu-kaolinite was more toxic than Cu-montmorillonite, probably because Cu was bound more strongly to montmorillonite or was partly concealed in the interlamellar spaces. Cu-kaolinite was even more toxic than Cu dissolved in a clay-free medium,

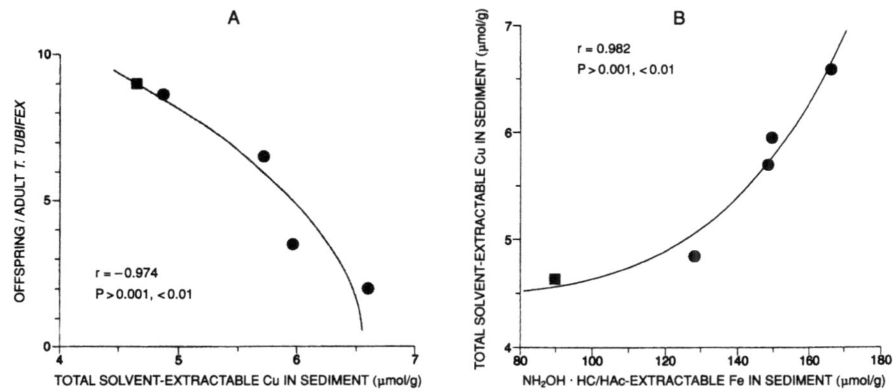

Fig. 5.30 A, B. Data from laboratory bioassays and sediment analyses showing apparent effects of FeOOH-bound Cu species on the reproduction of the benthic oligochaete (*Tubifex tubifex*) in sediment samples from a lake polluted with mine tailings (Larder Lake, Ontario, Canada). **A** Relationship between the offspring/adult ratio of *T. tubifex* and the sum of the sediment Cu fractions extracted sequentially with a series of solvents ($MgCl_2$, Na acetate/acetic acid, $NH_2OH \cdot HCl$/acetic acid, H_2O_2/HNO_3, and HNO_3/HCl, in that order); **B** relationship between the total solvent-extractable Cu and $NH_2OH \cdot HCl$/acetic acid (HAc)-extractable Fe concentrations in the sediments. (From Jackson et al. 1995)

indicating the importance of clay as a vehicle for Cu uptake. The toxicity of Cu-kaolinite was increased if the pH was raised from 7 to 8, presumably because the sorption capacity of the mineral was higher at pH 8. Luoma and Jenne (1977) found that the availability of ingested particle-bound metals for assimilation by a deposit-feeding clam depended on the characteristics of the metals as well as the particulate matter. Hardly any Zn or Co was taken up by the clam if the metal was bound to Fe or Mn oxide with which it had been coprecipitated; in contrast, the clam was able to extract Ag from Fe oxide. In another study, the toxicity of heavy metals (Cu, Zn, Ni, Cr, and Pb) in lake mud polluted with mine tailings was assessed by measuring adverse effects of mud samples of varying metal content on reproductive functions of the freshwater deposit-feeding oligochaete *Tubifex tubifex* (Jackson et al. 1995). Production of young appeared to be inhibited mainly by solvent-extractable forms of Cu bound to Fe (but not Mn) oxide (Fig. 5.30). The worms probably absorbed the Cu from the gut following ingestion of mud and release of Cu by solubilisation of Fe oxide during digestion. The respective contributions of the five metals to the observed toxic effect were estimated by regression analysis, which revealed inverse correlations between reproduction and solvent-extractable metal concentrations. Relative toxicity (as measured by the r-value for each metal) correlated almost perfectly with the sum of the first two ionisation potentials of the metal and was therefore greatest for Cu. Thus, Cu was by far the most toxic metal because it has the highest ionisation potentials, suggesting strong inhibition of enzymes by the formation of largely covalent bonds with their active sites. Unlike *T. tubifex*, the bottom-feeding fish species white sucker (*Catostomus commersoni*) seems to have difficulty releasing heavy metals from Fe oxides in sediments which it swallows (see above). In a group of lakes polluted by smelter fallout, the Cd content of white sucker liver was inversely related to the Fe oxide content of associated mud, suggesting that the oxide inhibited Cd uptake (Jackson et al. 1993). On the other hand, animals that consume aquatic insects may assimilate toxic metals sorbed to Fe oxide coatings on the exoskeletons of their prey (Hare et al. 1991). This mechanism of metal assimilation is analogous to that of *T. tubifex* even though it involves an entirely different pathway for the uptake of Fe oxides and associated metals.

5.6.2
Creation and Destruction of Toxic Trace Element Species

This topic has already been discussed in Section 5.5. As demonstrated by the examples mentioned there, the interconversion of relatively toxic and nontoxic trace element species by reactions with mineral colloids (whether direct effects like oxidation by MnOOH or indirect effects like suppression or promotion of microbial activities by clay) may be of considerable ecological importance.

5.7
Catalytic Activities of Minerals and Metal-Mineral Complexes

Oxides, clay minerals, clay–metal complexes, and metals in solution function as catalysts in many organic reactions (Malla et al. 1991; Sun Kou et al. 1992). The spontaneous abiotic decomposition of methyl Hg in the presence of Mn oxide could be an example of such a process (Jackson 1989; see Sect. 5.5). It seems doubtful whether nonbiological reactions like this are significant in natural environments within the biosphere, where chemical reactions are generally mediated by micro-organisms; but abiotic catalysis on colloid surfaces could conceivably occur in nature and has, in fact, been invoked as one of the mechanisms whereby humic matter is formed. Results of experiments performed under aseptic conditions led Wang and Huang (1987) to postulate that δ-MnO_2 (birnessite), a common soil mineral, catalyses the polycondensation of phenols and amino acids to form humic substances in soil. As Fe and Mn oxides sorb heavy metals, which are potential catalysts in their own right, the possibility arises that many varieties of catalytic functions are performed by different kinds of heavy metals sorbed to various oxide minerals. The result could be a great diversity of reactions catalysed by different minerals and by different sites on the same mineral, with a complex interplay between microbial activities and abiotic reactions. Such reactions could include transformation of toxic organic compounds into harmless products or into even more harmful ones.

Whatever their significance, if any, in natural environments at the present time, the catalytic functions of colloidal minerals are thought to have played a key role in the origin of life. Experiments on hypothetical pathways for prebiotic synthesis of biologically important macromolecules such as proteins have shown that clay minerals acting as catalysts and templates (i.e. rudimentary "enzymes") can mediate the polymerisation of amino acids and other building units of macromolecules and can even discriminate between the L- and D-optical isomers of amino acids (Jackson 1971a, b). It has been suggested that Al atoms of the edge faces of clay crystals take part in such reactions, and it is by no means improbable that heavy metals bound to these crystal faces by surface complexation contributed to the catalysis of prebiotic organic reactions. Literature on the possible importance of clay minerals in the origin of life has been reviewed by Theng (1974).

5.8 Practical Applications

5.8.1 Pollution Prevention and Abatement

5.8.1.1 *Disposal and Containment of Wastes*

It has long been standard practice to line waste disposal pits with clay to prevent toxic heavy metals and other pollutants from migrating into neighbouring soil, groundwater, and surface water (Förstner et al. 1991; Warith and Yong 1991) (see Chap. 6, this Vol.). Clay-lined pits have also been proposed for disposal of radioactive wastes (Brause et al. 1991; Davison et al. 1991; Czurda and Wagner 1991; see Chap. 2, this Vol.). The effectiveness of clay in containing metals is due to its low permeability and high sorption capacity; but to function properly, clay linings in disposal pits must be thick enough (Wagner 1991) and have to be kept moist to prevent the formation of cracks which might allow the escape of pollutants into the surrounding soil (Beszedits 1979).

It must be remembered, too, that some heavy metals are bound more strongly than others by mineral colloids, and that environmental variables play important roles. In a study of the diffusion of Cu, Zn, and Cd from an old tailings deposit downward into an illitic clay formation, Allard et al. (1991) observed that as the pH rose from 4.0 at the top of the clay formation (where it is in contact with tailings) to the range 5.5–6.0 at a depth of 40–50 cm, immobilisation of the metals decreased in the order Cu > Zn > Cd, suggesting that the degree of retardation was controlled by metal cation hydrolysis. Griffin et al. (1977) and Förstner et al. (1991) have pointed out that the usefulness of clay as a barrier at a waste disposal site is largely a function of the pH of the leachate. Higher pH values assist the sorption of cationic metals [e.g. Pb, Hg, Cd, Zn, Cu, and Cr (III)] but tend to prevent the sorption of anionic metals and metalloids [e.g. Cr (VI), As, and Se] (Griffin et al. 1977). Moreover, there has been some concern that Pb may be solubilised as an anionic species at high pH, as might be the case if alkaline fly ash is deposited in a disposal pit in soil (Majone et al. 1993; Roy et al. 1993). Nevertheless, Roy et al. (1993) reported that in such circumstances the anionic Pb is subject to sorption by kaolinite and illite; nor is the efficiency of sorption diminished by Cl^- or CO_3^{2-}.

The effectiveness of clay minerals in pollutant retention probably lies chiefly in their lack of permeability; as far as metal-binding is concerned, oxides and organic matter are more effective. For containment of metals at disposal sites on land a combination of clay minerals, organic matter, and oxides probably gives the most satisfactory results. Farquhar and Constable (1978) studied the fate of metals in landfill leachate percolating through soil and concluded that removal of metals (e.g. Zn) from soil solution became more efficient with increasing cation exchange capacity, which was mainly a function of the organic content of the soil. Most of the Zn, however, was not readily desorbed on coming into contact with water, suggesting that specific sorption rather than ordinary cation exchange was involved. In any event, organic matter was a more effective binding agent than clay. Fe and Mn oxides also immobilise metals at waste disposal sites (Förstner et al. 1991; Brown et al. 1994). Thus, Fe and Mn oxides precipitate spontaneously, removing metals from solution, when dredged sediment from an anoxic environment is exposed to free O_2 (Patrick et al. 1977). Nonetheless, oxidation, as we have seen, may lead to a net increase in the mobility of Cd and other metals in spite of partial immobilisation by oxides; consequently, subaqueous disposal of dredge spoil or mine tailings in a reducing environment where sulfides are generated is probably safer from the standpoint of minimising the solubilisation of metals (Förstner 1980; Khalid 1980). But for certain toxic elements such as As, an oxidising environment may be safer (see Sect. 5.5.1). In a reducing environment As bound to Fe oxide is released by reduction of the oxide and may be reduced from As(V) to As(III), thereby becoming more mobile and more toxic (Bowell 1994). If dredge spoil is buried in the ground, the most desirable soil conditions from the standpoint of minimising mobilisation of metals and uptake by plants include high pH and high concentrations of clay and organic matter (van Driel and Nijssen 1988).

Once heavy metal wastes have been buried at a land disposal site, the overlying ground surface is often revegetated (see Chapt. 6, this Vol.); but the plants may be inhibited by the metals, in which case clay may be used as a detoxifying agent. Experiments performed by Figge et al. (1995) showed that plant growth on mine tailings improved if a layer of expanding clay such as montmorillonite was deposited topically on the tailings. These results are in agreement with the research of Stotzky and others (see Sect. 5.6.1.2) demonstrating that montmorillonite can protect soil microbes from toxic metals.

Regarding disposal of wastes containing toxic organic compounds, or removal of such compounds from wastewaters or polluted environments, the possibility of using colloids or colloid-metal complexes that promote breakdown of the compounds to harmless products by abiotic catalysis or microbial activity, or both, is at least conceivable (see Sect. 5.7). This approach may be worth investigating.

5.8.1.2
Removal of Metals from Wastewater

It has been suggested that heavy metals be removed from industrial effluents by methods involving coprecipitation with Fe and Mn oxides or sorption by these oxides, or both (Lee 1975; Leckie et al. 1980; Oscarson et al. 1981c, 1983; Bailey et al. 1992; Sinicrope et al. 1992; Mott et al. 1993; Jackson and Bistricki 1995). Two of the proposed procedures involve utilisation of biogeochemical processes in natural ecosystems. Sinicrope et al. (1992) advocated removal of metals from wastewater by channeling the water in pulsed discharges twice a day through giant bulrush wetlands, where the metals would be scavenged and retained by precipitation of Fe oxide, thereby being prevented from contaminating environments downstream from the wetlands. Presumably the large quantities of organic matter and the anoxic conditions of the predominantly organic sediments in such environments would also play a key role in binding and retaining metals. A test of the feasibility of this method revealed that the efficiency of metal removal after one year was 84% for Pb, 75–78% for Cd, Zn and Cr, and only 55% for Ni. Obviously a certain degree of selectivity is involved, and it may be unavoidable. In any case, wetlands are commonly exploited for the purification of metal-bearing mine effluents. A different system based on the same general principles has been outlined by Jackson (1978) and Jackson and Bistricki (1995), who proposed that effluents from tailings ponds or other wastewaters charged with heavy metals be channeled through settling ponds fertilised with sewage to generate algal blooms, possibly with environmental conditions adjusted to favour the growth of the euglenophytes *Trachelomonas* and *Strombomonas*, which are enclosed in loricae encrusted with metal-scavenging Fe and Mn oxides. A system like this could immobilise metals efficiently owing to the combined effects of the following features: (1) high density of metal-binding planktonic organisms and nonliving particles, many of them with oxide coatings, which are available throughout the growing season to withdraw metals from the water by sorption and bio-accumulation, conveying them to the bottom on settling out; (2) creation of alkaline conditions in the water by intense photosynthesis, improving the efficiency of sorption; and (3) production of H_2S and other sulfides, and creation of alkaline conditions on the bottom by putrefaction of algal remains accompanied by bacterial reduction of SO_4^{2-} (a major constituent of acid mine drainage). The metals would be strongly immobilised by the sulfides, although the efficiency of entrapment of metals in the bottom of the pond would probably vary systematically with metal properties (Jackson 1978, 1979). This method would have the advantage of being both economical and effective.

The use of clay minerals to remove metals from effluents has also been considered. Sharma et al. (1991) and Gupta et al. (1992) suggested the use of "china clay" (kaolin) for scavenging metals (e.g. Cr dyes, Ni, and Zn) from industrial wastewater. A possible advantage of this material is that it is cheap and abundant (Sharma et al. 1991), although large deposits of it occur only in certain regions of the world. In addition to natural clay, certain specially modified clay minerals have been tried out experimentally. For instance, de Boodt (1991) examined possible applications of heat-treated clay minerals with artificially added $Al(OH)_3$ coatings and concluded that the product is an efficient, economical sorbent for removing both cationic and anionic metal species from wastewater. Cadena et al. (1990) developed techniques for removal of metals from solution by "tailored bentonite" (bentonite complexed with tetramethylammonium [TMA^+] ions). The effects of the clay-organic complex, however, were selective; his data show that addition of TMA^+ to the clay enhanced the sorption of Pb but had little effect on Cr.

In all methods of wastewater purification based on the use of a metal-scavenging additive like clay or Fe oxide, thought must be given not only to the removal but also to the disposal and ultimate fate of the metals and the possibility that they may at some future time be remobilised, causing further trouble. Improper disposal could lead to transfer of the pollution problem to another locality rather than a real solution of the problem. Possible regeneration and reuse of metal-scavenging substances should also be considered.

5.8.1.3
Remediation of Polluted Environments

Many rivers, lakes, river–lake systems, and near-shore marine environments are polluted with heavy metals, which are concentrated in fine-grained bottom sediments and suspended matter. In a river system the contamination may be spread over great distances by fluvial transport, as observed, for instance, in the Rhine River and the previously discussed Hg-polluted river systems in Canada. Once a natural aquatic system has been contaminated, what remedial measures, if any, are possible? This is a difficult and complex problem, as all options entail certain drawbacks and limitations involving questions of technical feasibility, cost, undesirable side effects, and limited effectiveness (Jackson et al. 1980b; Jackson and Woychuk 1980a, b). But even if nothing is done, a contaminated system will recover gradually (provided there is no further pollution) as the contaminated sediments are eroded and dispersed or buried by uncontaminated material. This brings us to the possibility that clay may play a helpful role, raising the question of whether it is feasible and desirable to accelerate natural recovery by deliberately introducing large quantities of uncontaminated mineral particles (e.g. clay and silt) into a polluted system.

Several workers have discussed the possibility that the Hg-contaminated Wabigoon–English–Winnipeg River system and its riverine lakes (see Sect. 5.4.4.3) can be ameliorated by introducing masses of uncontaminated clay and silt into the polluted system and that similar procedures can be employed elsewhere (Jackson et al., 1980b; Jackson and Woychuk 1980a, b; Rudd and Turner 1983; Parks and Hamilton 1987). In theory, mineral detritus dumped into the river system at certain localities would be transported through the watercourse by normal fluvial action and laid down in basins of deposition, covering the contaminated surficial sediments laid down earlier, besides diluting contaminated material and inhibiting microbial methylating activity. The idea is not without merit. There are grounds for believing that a layer of clean sediment, even if only a few millimeters thick, would bring about a marked improvement (Parks and Hamilton 1987), and there is empirical evidence that large-scale intrusion of suspended clay and silt into a river system or lake depresses methyl Hg production and the accumulation of Hg by fish (Rudd and Turner 1983; Jackson 1988b, 1991, 1993b, c). Besides, there are some local sources of clay and silt along the banks of the river and, especially, in a large unpolluted lake in the system itself, just upstream from the source of pollution. The proposed method has the additional advantage of employing harmless natural materials and making use of a completely natural processes, minimising harmful side effects. However, this remediation scheme, like any other, has limitations and serious disadvantages which cast doubt on the advisability of implementing it (Jackson et al. 1980; Jackson and Woychuk 1980a, b, 1981). In the first place, the method would probably not be very successful unless preceded by wholesale dredging of the thick deposit of decomposing wood chips blanketing the riverbed immediately downstream from the source of pollution, as disruption by gas bubbles, resuspension by bottom scour, and infiltration into the coarse organic debris would almost certainly prevent deposition of a coherent clay-silt layer on top of this material. Moreover, Hg resuspended or solubilised from the organic deposit is continually contaminating the waters downstream from it, and this would partially cancel any beneficial effect of sediment remediation in the more distal regions. Even if contaminated lake and river sediments further down the system could be largely covered with uncontaminated detritus, buried Hg could be brought back into the aquatic environment by bottom erosion and the burrowing action of benthic animals. Dredging followed by the transfer of massive quantities of clay and silt into the river from unpolluted lacustrine and terrestrial deposits would inevitably involve substantial expense and environmental damage, and the likelihood of achieving sufficient success to justify these drawbacks is doubtful. Another problem is the sheer magnitude of the river system. The distance from the source of pollution to the south basin of Lake Winnipeg, into which the system drains, is more than 160 km, and there are many riverine lakes along the way. Besides, in this system, as in others studied by the author, methyl Hg production is a function of environmental parameters rather than total Hg and may even vary inversely with total Hg (Jackson and Woychuk 1980a, b; Jackson et al. 1982); partial burial of sedimentary Hg, therefore, is no guarantee of a proportional decrease in methyl Hg production. The best plan of all may be to leave the system alone and let Nature take her course.

One possible approach to remediation of metal-polluted soil or sediment is large-scale extraction of the metals using special techniques. Hamed et al. (1983) demonstrated experimentally that 75–95% of the Pb(II) sorbed to kaolinite can be removed by electrokinesis. They concluded from this that electrokinetic techniques can be adapted to the reclamation of soils polluted with heavy metals, including radionuclides, and that other applications such as tests for detection of leaks in clay linings of disposal pits might also be possible. More extensive research, however, is needed to determine (1) how effective this method is in the removal of metals from soil components that bind metals more strongly than kaolinite does, and (2) whether the method is truly practical for wholesale in situ removal of metals from natural soils.

Another remediation strategy that has been proposed is in situ conversion of toxic elements to less toxic forms through environmental manipulations such as the addition of Mn oxyhydroxide to promote oxidation of As (III) to As (V) (Oscarson et al. 1983). Simple laboratory experiments have shown that this is possible, but whether it is practicable and desirable as a method of environmental amelioration in a complex natural system is uncertain.

5.8.2
Pest Control

The work of Al-Sabri et al. (1993) suggests that clay minerals with sorbed Cu could be used to exterminate the freshwater snails which carry the parasite that causes schistosomiasis, a serious, widespread disease of tropical countries. The Cu is the agent that actually kills the snails, but clay promotes ingestion of Cu because the snails are filter feeders (see Sect. 5.6.1.2).

5.9
Specialised Research Techniques

A number of powerful special techniques are valuable for in-depth research on interactions between trace elements, colloids, and organisms. The more advanced methods permit direct study of sorbed metals, and they yield detailed, precise, refined, and very specific results, such as analytical data for individual microscopic particles in heterogeneous mixtures, and information on the immediate environment of a metal (e.g. the coordination of ligands around sorbed metal ions) and the type of bonding involved. Moreover, sundry *combinations* of techniques, including conventional methods as well as the more sophisticated ones, are particularly useful, as they provide different kinds of relevant information, making possible greater depth and breadth of understanding and more meaningful interpretations (Pickering 1979; Leppard et al. 1988, 1989; McBride 1991; Jackson and Bistricki 1995); and the opportunities for advancement of knowledge are greatly multiplied if samples representing effects of different variables, such as natural environmental factors or experimental treatments, are compared. There are limitless possibilities for creative use of combinations of methods, sampling strategies, and experimental designs.

It also has to be said that advanced modern techniques have serious limitations. One problem is that the methods are not readily accessible to many scientists because they entail heavy expenses and are time-consuming, highly specialised, and difficult. In addition, there are technical limitations and pitfalls; even gross errors may be made through inadvertent production of artefacts or erroneous results, as may occur, for instance, if samples are not properly prepared for analysis. Nevertheless, the potential rewards in terms of scientific achievement are enormous. The following examples demonstrate what can be accomplished with the sophisticated technological aids currently available:

1. Electron spin resonance (ESR) spectroscopy has shown that exchangeable metal cations of clay minerals are strongly hydrated (i.e. consist of species such as $Cu(H_2O)_6^{2+}$) and are therefore held at the mineral surface by electrostatic attraction rather than direct bonding by surface ligands (Pickering 1979; McBride 1991b). ESR data have also been used in experiments to determine whether metals sorbed to clay are bound mainly by edge sites or 001 sites, although Kinniburgh and Jackson (1981) have rightly warned that such data may be highly misleading unless natural oxide coatings and other surface impurities are removed beforehand. In the characterisation of Cu sorbed to montmorillonites and kaolinites, ESR can distinguish between octahedral Cu and the surface-sorbed and interlamellar fractions combined, but it cannot differentiate between the surface-sorbed and interlamellar fractions (Mosser et al. 1992).
2. X-ray photoelectron spectroscopy (XPS), which is sensitive to local ordering around cations, has been used to characterise the bonding state of Cu(II) sorbed to montmorillonite and kaolinite (Mosser et al. 1992). Using XPS, it was possible to differentiate between Cu in octahedral, surface-sorbed, and interlamellar positions. In this regard XPS was able to provide more exact information than ESR. Using XPS to

examine the bonding of Co(II) and Ni(II) by montmorillonite and hectorite, Davison et al. (1991) were able to distinguish between labile and less readily desorbed fractions, and they concluded that sorbed Co^{2+} and Ni^{2+} both showed signs of being in a common "MO_6" (i.e. octahedral) ligand environment.

3. The isotope-exchange technique distinguishes between metals of the "labile pool," which are loosely sorbed (and hence isotopically exchangeable), and metals that are strongly – even irreversibly – bound (and hence are not isotopically exchangeable) (McBride 1991b).
4. Mössbauer spectroscopy provides information on the immediate atomic environment of Fe and can be used for the identification and study of different kinds of Fe oxide, including "amorphous" gels. Xing et al. (1995) employed the Mössbauer technique to investigate sorption of metals by amorphous Fe oxide in soil, and the spectra were helpful in determining the presence or absence of Fe oxide after selective solvent extraction.
5. Energy dispersive X-ray microanalysis in conjunction with electron microscopy has been used to determine the elemental composition of microscopic components of heterogeneous mixtures of particulate matter, such as plankton and sediments, by analysing them singly, one by one (Leppard et al. 1988, 1989, 1990; Leppard 1992; Jackson and Bistricki 1995). With this technique, visually identifiable or describable objects in such a mixture (e.g. a single bacterial cell), and even different components of a single object (e.g. a cell wall or an organelle or inclusion in the cell's cytoplasm), can be analysed *individually*. The method has also been employed to analyse distinct microscopic domains within various solids (e.g. separate Mn oxide and Fe oxide zones in marine Mn-Fe nodules) (Burns and Fuerstenau 1966; Calvert and Price 1970; Stiff et al. 1982). A single analysis yields a spectrum that instantly provides semiquantitative estimates of many metallic and nonmetallic elements. Furthermore, mineral particles can be identified by electron diffraction. With the use of special stains and specific markers, the method can also be adapted to the identification of certain kinds of organic compound (e.g. proteins), and it is now possible to perform analyses on wet samples in their native state using state-of-the-art equipment such as the ElectroScan 2020 "environmental" scanning electron microscope (G.G. Leppard, pers. comm.). Microanalysis represents an immense quantum leap beyond the much more limited methods of conventional bulk analysis. This unique, refined, and powerful technique has the potential to revolutionise biogeochemistry by furnishing a wealth of information that would be impossible to obtain otherwise. However, it is subject to certain limitations and

problems that must be kept clearly in mind. Limited sensitivity is one obstacle to full realisation of the method's potential, as elements in minute particles may be undetectable unless there is considerable enrichment; and if an element is concentrated in one localised region of a particle (e.g. an inclusion in a single-celled organism), it might be missed. The plankton analysed by Jackson and Bistricki (1995) came from lakes severely polluted with Cu, Zn, and Cd, but Cd was not detected because of its relatively low abundance. (Increasing the time of exposure of each sample to electron irradiation might help to solve such a problem, but it would mean a substantial increase in time and cost and might result in destruction of the samples.) Another limitation is that certain peaks in microanalysis spectra coincide or overlap; thus, the peaks for Hg and S practically coincide, making determination of Hg difficult. A particularly serious problem is that alteration of a sample through the use of wrong procedures for sample preservation, storage, and pretreatment may result in creation of artefacts and production of unreliable data. Thus, proper handling of samples is of the essence (Leppard et al. 1989, 1990).

5.10 Concluding Remarks: Some Thoughts About Future Research

The biogeochemical and ecological effects of colloidal minerals in natural environments are highly variable and complex, and in-depth investigation of them is correspondingly difficult. In soil and aquatic ecosystems, different varieties and quantities of minerals, trace elements, organic and inorganic ligands, and organisms are mixed together under spatially and temporally varying physicochemical conditions, and they interact with each other in a multitude of direct and indirect ways (Jackson 1989, 1995). Many phenomena occur simultaneously, some reinforcing and others tending to cancel each other's effects, and the *net* effect may not be predictable or readily comprehensible. What, then, should be the guiding principles that influence the course of future research in this field? The following considerations may be helpful.

Interdisciplinary Studies. The value of interdisciplinary research and knowledge can scarcely be overstated. The biogeochemical and ecological phenomena examined here overlap many conventional disciplines in both the physical and the biological sciences. To see the subject as a whole, it is necessary to transcend artificial boundaries between different specialties. Ideally, research in this field should represent a marriage of breadth and depth.

Integration of Field and Laboratory Research. A logical approach is to combine the analysis of field samples with controlled experiments, as the two types of study are complementary, and neither by itself is fully satisfactory (Pickering 1979; Swift and McLaren 1991; Jackson 1995). Experiments can yield important information on processes occurring under defined conditions and can greatly assist the interpretation of data obtained from field samples; but, as experimental systems are simplified, more or less artificial representations of natural environments, they afford, at best, only limited insight into phenomena occurring under natural conditions. Experimental data alone can lead to distorted, oversimplified conceptions of natural phenomena, and even gross errors (as in experiments in which impurities associated with clay minerals are not taken into account). Of course, experimental systems vary widely in the degree to which they approximate natural environments (Jackson 1995). Empirical field studies offer the advantage of providing information on phenomena actually occurring under natural conditions; but, because natural environments are so complex and involve so many interrelated variables and phenomena, the results are often too ambiguous to support firm conclusions. An observed relationship or effect may be the *net* result of many different processes, and a correlation is not necessarily evidence for cause and effect. Without additional information from experiments, it may not be possible to separate the various factors contributing to the observed result. A typical example has been provided by Chen et al. (1991), who presented field data revealing that the Cd- and Cu-sorption capacities of suspended matter in major Chinese rivers vary from north to south because of variation in particle composition. According to their data, the sorption capacities increase with clay content, organic content, cation exchange capacity, montmorillonite/kaolinite ratio, and humic acid/fulvic acid ratio, and are inversely related to the abundance of Fe oxide and humic matter! Their results are interesting and suggestive but are too ambiguous, by themselves, to support definite conclusions about cause and effect.

Comparative Studies. Comparative studies may yield far more information than the investigation of a single narrowly defined subject. Examples include comparisons between data representing (1) different environments (e.g. different sites along an environmental gradient, different lakes in a given field area, and different sites within a single lake or river); (2) a single field site sampled at different times of year or in different years; (3) different kinds of trace elements; (4) effects of different coexisting sorbents and complexing agents in natural environments; and (5) different kinds of organisms in a natural community, or different stages in the life of a particular kind of organism.

Integration of Theoretical and Empirical Knowledge. Attempts to fit Nature into an oversimplified, or even inappropriate, Procrustean theoretical framework may lead to inadequate or erroneous conceptions about phenomena occurring in natural ecosystems. Common examples include (1) the use of thermodynamic calculations and phase diagrams to describe or predict geochemical reactions, with too little attention to kinetics and mechanisms, which often have the dominant effects, at least in the short term; (2) reliance on principles of pure inorganic chemistry to explain reactions that are actually mediated by organic (e.g. microbial) activities; (3) use of theoretical calculations to predict metal speciation even though the actual forms of the metals may be altogether different owing to effects of complexing agents and organisms; (4) the use of "models" to describe or predict natural phenomena (an approach that is often unsuccessful, or only partially successful, because it is based on faulty premises or incorrect estimates, or because important factors have not been taken into account); and (5) interpretation of natural processes in terms of physicochemical parameters of the bulk environment, overlooking effects of microenvironments. Nonetheless, judicious application of basic physicochemical theory is necessary and has, in many cases, been very helpful, as in studies revealing relationships between properties of heavy metals and empirical data on the behaviour of these metals in nature (Jackson 1978, 1979; Jackson et al. 1980a, 1995). It is hardly necessary to add that theory must always be verified by empirical observation.

Technological Aids. In future, the most fundamental and innovative research will depend heavily on continued development, and more extensive application, of powerful, sophisticated analytical methods and other technological advances, such as the ones discussed in Section 5.9. One of the great challenges will be to devise new techniques for the noninvasive direct examination of colloid–trace element–organism interactions in situ under natural conditions. Of particular value would be the invention of more precise, powerful, and sensitive instruments for detailed elemental and organic microanalysis of microscopic objects such as clay crystals and bacterial cells in complex natural materials such as soils and sediments, and in water samples, in their native state without altering them or disturbing their mutual spatial relations.

Areas of Inquiry. There is clearly a need for more basic and applied research on interactions between trace elements, colloidal minerals, complexing agents, and organisms under different environmental conditions, with reference to the biological effects and biogeochemical pathways of the elements, and with emphasis on phenomena actually occurring in natural ecosystems. Some examples of important interrelated areas of inquiry are as follows:

1. Trace element speciation (still a poorly known subject), and binding and release of trace element species by different sorbents and ligands, under different environmental conditions in various natural ecosystems. Direct characterisation of different forms of the elements. Partitioning of elements among different binding agents. Effects of organisms (e.g. microbes) as well as physicochemical variables on these processes.
2. Ecological effects of trace element speciation, and binding and release of the elements by different coexisting sorbents and complexing agents, under different conditions: implications for the bio-availability, bio-accumulation, and various direct and indirect biological effects of the trace elements, with reference to different kinds of organisms in communities of different natural ecosystems under varying conditions. Impact of elements, colloids and complexing agents on entire communities (as distinct from isolated species used in bioassays), including effects on interactions between different organisms (mutualism, predator-prey relations, competition, release of nutrients to primary producers by microbes, etc.) as well as direct effects on particular organisms.
3. Effects of coatings and other surface impurities on clay and oxide particles (a crucially important area of inquiry which has been largely neglected, even though lip service has been paid to it repeatedly in the literature).
4. Selective effects due to specific metal and colloid properties, biochemical activities, and environmental factors (e.g. preferential sorption and catalysis at mineral surfaces, template effects, specific organic–inorganic interactions, and other selective structure–function relations, and specific effects of metals, colloids, complexing agents, and interactions between them on particular kinds of organisms and on particular biological functions and biochemical reactions – as, for instance, in selective protection of particular organisms by particular colloids, and selective biological effects of particular chemical species). Specific physiological and biochemical adaptations of organisms to trace elements, and effects of colloids, complexing agents, and environmental variables.
5. The impact of colloids on microbial activities affecting heavy metal speciation, mobility, and bio-availability, and the role of colloids, trace elements, and other factors in microbial ecology generally. (Even now, microbial ecology is still in its infancy owing to the tech-

nical difficulty of studying the nature and activities of in situ microbial populations in natural environments (Jackson 1995). There is a need for more sophisticated analysis of the many interrelated phenomena connected with metal–colloid–microbe interactions as they actually occur in nature.)

6. Applications of microanalysis techniques to the study of various interacting trace elements, colloids, complexing agents, and microscopic organisms under different conditions in complex, heterogeneous natural systems.
7. Phenomena occurring in microenvironments created by organisms and colloids (e.g. a chemically distinctive zone surrounding a bacterial micro-colony in soil, or the surface of a colloidal mineral particle).
8. Mechanisms of interactions between trace element species, colloids, complexing agents, and organisms at the molecular and atomic level.
9. Effects of mixtures of interacting trace elements, colloids, complexing agents, and organisms under different conditions (e.g. synergistic and antagonistic effects).
10. Design and application of safe, effective, and economical methods for pollution prevention (e.g. removal of metals from wastewater), remediation of polluted environments (e.g. immobilisation and detoxification of toxic elements), and waste disposal. Study of the most effective combinations of biological and physicochemical processes.

It is worth mentioning that in any study of processes occurring in natural environments it is very important to collect a wide range of background data (both physicochemical and biological) which are likely to be relevant. One of the commonest deficiencies of papers on environmental phenomena is a lack of enough auxiliary information to interpret and explain the principal results.

In conclusion, future progress in our understanding of interactions between metals, colloids, complexing agents, and organisms, and elucidation of their biogeochemical and ecological significance, will require the application of various powerful modern analytical techniques, a wide range of techniques and information, integration of field studies and controlled experiments, further technical innovations, integration of knowledge, concepts, and methods from different fields in both the biological and the physical sciences, and the depth of insight and breadth of knowledge and vision needed to achieve an interdisciplinary synthesis of a broad spectrum of information and ideas.

Acknowledgements. The financial support for this study came from the government of Canada (the Department of the Environment, otherwise known as Environment Canada). I thank all publishers and authors who kindly permitted me to use previously published illustrations.

References

Abd-Elfattah A, Wada K (1981) Adsorption of lead, copper, zinc, cobalt, and cadmium by soils that differ in cation-exchange materials. J Soil Sci 32:271–283
Aiken GR, McKnight DM, Wershaw RL, MacCarthy P (eds) (1985) Humic substances in soil, sediment, and water. John Wiley (Wiley-Interscience), New York, 692 pp
Allard B, Håkansson K, Karlsson S, Sigas E (1991) A field study of diffusion controlled migration of copper, zinc, and cadmium in a clay formation. Water Air Soil Pollut 57–58:259–268
Al-Sabri IYN, Smith JD, Thomas JD (1993) Copper molluscicides for control of schistosomiasis. 3. Adsorption by clay suspensions. Environ Sci Technol 27: 299–303
Anderson MA (1981) Kinetic and equilibrium control of interfacial reactions involving inorganic ionic solutes and hydrous oxide solids. In: Brinckmann FE, Fish RH (eds) Environmental speciation and monitoring needs for trace metal containing substances from energy-related processes. US Dept of Commerce, Washington, DC, pp 146–161
Andersson A (1977) Heavy metals in Swedish soils: on their retention, distribution and amounts. Swed J Agric Res 7:7–20
Andersson A (1979) Mercury in soils. In: Nriagu JO (ed) The biogeochemistry of mercury in the environment. Elsevier/North-Holland Biomedical Press, Amsterdam, pp 79–112
Andren AW, Nriagu JO (1979) The global cycle of mercury. In: Nriagu JO (ed) The biogeochemistry of mercury in the environment. Elsevier/North-Holland Biomedical Press, Amsterdam, pp 1–21
Atanassova ID (1994) Adsorption and desorption of Cu at high equilibrium concentrations by soil and clay samples from Bulgaria. Environ Pollut 87:17–21
Baas Becking LGM, Kaplan IR, Moore D (1960) Limits of the natural environment in terms of pH and oxidation-reduction potentials. J Geol 68:243–284
Babich H, Stotzky G (1977a) Effect of cadmium on fungi and on interactions between fungi and bacteria in soil: influence of clay minerals and pH. Appl Environ Microbiol 33:1059–1066
Babich H, Stotzky G (1977b) Reductions in the toxicity of cadmium to microorganisms by clay minerals. Appl Environ Microbiol 33:696–705
Babich H, Stotzky G (1979) Abiotic factors affecting the toxicity of lead to fungi. Appl Environ Microbiol 38:506–513
Babich H, Stotzky G (1982) Toxicity of nickel to microorganisms in soil: influence of some physicochemical characteristics. Environ Pollut Ser A 29:303–315
Babich H, Stotzky G (1983a) Influence of chemical speciation on the toxicity of heavy metals to the microbiota. In: Nriagu JO (ed) Aquatic toxicology. John Wiley, New York, pp 1–46

Babich H, Stotzky G (1983b) Physicochemical factors of natural reservoirs affect the transformation and exchange of heavy metals toxic to microbes. Ecol Bull (Stockholm) 35:315–323
Babich H, Stotzky G (1983c) Temperature, pH, salinity, hardness, and particulates mediate nickel toxicity to eubacteria, an actinomycete, and yeasts in lake, simulated estuarine, and sea waters. Aquat Toxicol 3:195–208
Baham J, Sposito G (1994) Adsorption of dissolved organic carbon extracted from sewage sludge on montmorillonite and kaolinite in the presence of metal ions. J Environ Qual 23:147–153
Bailey RP, Bennett T, Benjamin MM (1992) Sorption onto [sic] and recovery of Cr(VI) using iron oxide coated sand. Water Sci Technol 26:1239–1244
Barbeau K, Moffett JW, Caron DA, Croot PL, Erdner DL (1996) Role of protozoan grazing in relieving iron limitation of phytoplankton. Nature 380:61–64
Barnes LSD, Walne PL, Dunlap JR (1986) Cytological and taxonomic studies of euglenales. I. Ultrastructure and envelope elemental composition in *Trachelomonas*. Brit Phycol J 21:387–397
Barrie LA, Gregor D, Hargrave B, Lake R, Muir D, Shearer R, Tracey B, Bidleman T (1992) Arctic contaminants: sources, occurrence and pathways. Sci Total Environ 122:1–74
Barrow NJ, Bowden JW, Posner AM, Quirk JP (1981) Describing the adsorption of copper, zinc and lead on a variable charge mineral surface. Australian J Soil Res 19:309–321
Belamie R, Gouy V (1992) Introduction des pollutants dans le milieu fluvial. Influence du ruissellement des sols. Océanis 18:505–521
Belzile N, Tessier A (1990) Interactions between arsenic and iron oxyhydroxides in lacustrine sediments. Geochim Cosmochim Acta 54:103–109
Beneš P, Havlík B (1979) Speciation of mercury in natural waters. In: Nriagu JO (ed) The biogeochemistry of mercury in the environment. Elsevier/North-Holland Biomedical Press, Amsterdam, pp 175–202
Benjamin MM, Leckie JO (1981) Multiple-site adsorption of Cd, Cu, Zn, and Pb on amorphous iron oxyhydroxide. J Colloid Interface Sci 79:209–221
Berrang PG, Grill EV (1974) The effect of manganese oxide scavenging on molybdenum in Saanich Inlet, British Columbia. Mar Chem 2:125–148
Berrow ML, Burridge JC (1991) Uptake, distribution, and effects of metal compounds on plants. In: Merian E (ed) Metals and their compounds in the environment. VCH, Weinheim, pp 399–410
Beszedits S (1979) Mercury removal from effluents and wastewaters. In: Nriagu JO (ed) The biogeochemistry of mercury in the environment. Elsevier/North-Holland Biomedical Press, Amsterdam, pp 231–276
Bewley RJF, Stotzky G (1983a) Effects of cadmium and zinc on microbial activity in soil; influence of clay minerals. Part I: metals added individually. Sci Total Environ 31:41–55
Bewley RJF, Stotzky G (1983b) Effects of cadmium and zinc on microbial activity in soil; influence of clay minerals. Part II: metals added simultaneously. Sci Total Environ 31:57–69
Bishop JN, Neary BP (1976) Mercury levels in fish from northwestern Ontario, 1970–1975. Ontario Ministry of the Environment, Toronto, Ontario, Canada
Bolland MDA, Posner AM, Quirk JP (1977) Zinc adsorption by goethite in the absence and presence of phosphate. Australian J Soil Res 15:279–286

de Boodt MF (1991) Application of the sorption theory to eliminate heavy metals from waste waters and contaminated soils. In: Bolt GH, de Boodt MF, Hayes MHB, McBride MB, de Strooper EBA (eds) Interactions at the soil colloid-soil solution interface. Kluwer, Dordrecht, pp 293-320
Bothner MHD, Carpenter R (1973) Sorption-desorption reactions of mercury with suspended matter in the Columbia River. Proc Symp Radioactive Contamination of the Marine Environment, 1972, Vienna (IAEA-SM-158/5), pp 73-87
Boust D, Joron JL (1986) Geochemical behaviour of Eu-152, Am-241, and stable Eu in oxic abyssal sediments. In: Bulman RA, Cooper JR (eds) Speciation of fission and activation products in the environment. Elsevier Applied Science, London, pp 334-338
Bowell RJ (1994) Sorption of arsenic by iron oxides and oxyhydroxides in soils. Appl Geochem 9:279-286
Bowen HJM (1966) Trace elements in biochemistry. Academic Press, London
Boyle EA (1979) Copper in natural waters. In: Nriagu JO (ed) Copper in the environment, part I (ecological cycling). John Wiley, New York, pp 77-88
Brause CJC, Morris DE, Eller PG, Buscher TC, Conradson SE (1991) Changes in U(VI) speciation upon sorption onto montmorillonite from aqueous and organic solutions [abstract]. Program and Abstracts, 28th Annu Meeting of Clay Minerals Society, 5-10 Oct 1991, Houston, Texas. Lunar & Planetary Inst (LPI), Houston, Texas, p 30 (LPI Contrib no 773)
Brown DA, Kamineni DC, Sawicki JA, Beveridge TJ (1994) Minerals associated with biofilms occurring on exposed rock in a granitic underground research laboratory. Appl Environ Microbiol 60:3182-3191
Bruemmer GW, Gerth J, Herms U (1986) Heavy metal species, mobility and availability in soils. Z Pflanzenernähr Bodenkd 149:382-398
Brümmer G, Tiller KG, Herms U, Clayton PM (1983) Adsorption-desorption and/or precipitation-dissolution processes of zinc in soils. Geoderma 31:337-354
Brümmer GW (1986) Heavy metal species, mobility and availability in soils. In: Bernhard M, Brinckman FE, Sadler PJ (eds) The importance of chemical "speciation" in environmental processes. Springer, Berlin Heidelberg New York, pp 169-192
Buffle J, Perret D, Newman M (1992) The use of filtration and ultrafiltration for size fractionation of aquatic particles, colloids, and macromolecules. In: Buffle J, van Leeuwen HP (eds) Environmental particles, vol 1. Lewis, Boca Raton, pp 171-230
Burns RG, Fuerstenau DW (1966) Electron-probe determination of inter-element relationships in manganese nodules. Am Mineral 51:895-902
Cadena F, Rizvi R, Peters RW (1990) Feasibility studies for the removal of heavy metals from solution using tailored bentonite. In: Martin JP, Cheng S-C, Susavidge MA (eds) Hazardous and industrial wastes. Technomic Publishing Co, Lancaster, Pennsylvania, pp 77-94
Cain DJ, Luoma SN, Carter JL, Fend SV (1992) Aquatic insects as bioindicators of trace element contamination in cobble-bottom rivers and streams. Can J Fish Aquat Sci 49:2141-2154
Calder LM (1988) Chromium contamination of groundwater. In: Nriagu JO, Nieboer E (eds) Chromium in the natural and human environments. John Wiley, New York, pp 215-229
Calvert SE, Price NB (1970) Composition of manganese nodules and manganese carbonates from Loch Fyne, Scotland. Contrib Mineral Petrol 29:215-233
Calvert SE, Price NB (1977) Geochemical variation in ferromanganese nodules and associated sediments from the Pacific Ocean. Mar Chem 5:43-74

Candelone J-P, Hong S, Boutron C (1993) The changing occurrence of Pb, Cd, Zn, and Cu in Greenland snow from 1772 to 1965. In: Allan RJ, Nriagu, JO (eds) Heavy metals in the environment (Proc Int Conf, Sept 1993, Toronto). CEP Consultants, Edinburgh, Scotland, vol 2, pp 575–577
Chen JS, Cheng CQ, Wang FY (1991) Geographical tendency of geochemical characteristics and binding capacity of suspended matter from the five main rivers of east China for heavy metals. In: Vernet J-P (ed) Heavy metals in the environment. Elsevier, Amsterdam, pp 125–135
Chen J, Tang F, Wang F (1995) Mobilization of mercury from estuarine suspended particulate matter: a case study in the Yalujiang estuary, northeast China. Water Qual Res J Can 30:25–32
Collins YE, Stotzky G (1989) Factors affecting the toxicity of heavy metals to microbes. In: Beveridge TJ, Doyle RJ (eds) Metal ions and bacteria. John Wiley, New York, pp 31–90
Collins YE, Stotzky G (1992) Heavy metals alter the electrokinetic properties of bacteria, yeasts, and clay minerals. Appl Environ Microbiol 58:1592–1600
Cowan CE, Zachara JM, Smith SC, Resch CT (1992) Individual sorbent contributions to cadmium sorption in ultisols of mixed mineralogy. Soil Sci Soc Am J 56:1084–1094
Cranston RE, Buckley DE (1972) Mercury pathways in a river and estuary. Environ Sci Technol 6:274–278
Czurda KA, Wagner J-F (1991) Cation transport and retardation processes in view of the toxic waste deposition problem in clay rocks and clay liner encapsulation. Engin Geol 30:103–113
Davis JA, Leckie JO (1978) Effect of adsorbed complexing ligands on trace element uptake by hydrous oxides. Environ Sci Technol 12:1309–1315
Davis-Carter JG, Shuman LM (1993) Influence of texture and pH of kaolinitic soils on zinc fractions and zinc uptake by peanuts. Soil Sci 155:376–384
Davison N, McWhinnie WR, Hooper A (1991) X-ray photoelectron spectroscopic study of cobalt(II) and nickel(II) sorbed on hectorite and montmorillonite. Clays Clay Miner 39:22–27
Debosz K, Babich H, Stotzky G (1985) Toxicity of lead to soil respiration: mediation by clay minerals, humic acids, and compost. Bull Environ Contam Toxicol 35:517–524
Degens ET (1965) Geochemistry of sediments. Prentice-Hall, Englewood Cliffs, New Jersey, 342 pp
De Vitre R, Belzile N, Tessier A (1991) Speciation and adsorption of arsenic on diagenetic iron oxyhydroxides. Limnol Oceanogr 36:1480–1485
D'Itri FM (1991) Mercury contamination – what we have learned since Minamata. Environ Monit Assess 19:165–182
Doelman P (1978) Lead and terrestrial microbiota. In: Nriagu JO (ed) The biogeochemistry of lead in the environment. Elsevier/North-Holland Biomedical Press, Amsterdam, pp 343–353
Douglas BE, McDaniel DH (1965) Concepts and models of inorganic chemistry. Blaisdell Publishing, Waltham, Massachusetts, 510 pp
Du W (1992) The sorption of P related to the adsorption of Cu and Zn [abstract]. Can J Soil Sci 72:306
Duarte AC. Pereira ME, Oliveira JP, Hall A (1991) Mercury desorption from contaminated sediments. Water Air Soil Pollut 56:77–82

Dunlap JR, Walne PL (1985) Fine structure and biomineralization of the mucilage in envelopes of *Trachelomonas lefevrei* (Euglenophyceae). J Protozool 32:437–441
Dunlap JR, Walne PL (1987) Variations in envelope morphology and mineralization in *Trachelomonas lefevrei* (Euglenophyceae). J Phycol 23:556–564
Dunlap JR, Walne PL, Bentley J (1983) Microarchitecture and elemental spatial segregation of envelopes of *Trachelomonas lefevrei* (Euglenophyceae). Protoplasma 117:97–106
Dynes JJ, Huang PM (1995) Influence of citrate on selenite sorption–desorption on short-range ordered aluminum hydroxides. In: Huang PM, Berthelin J, Bollag J-M, McGill WB, Page AL (eds) Environmental impact of soil component interactions. Lewis (CRC), Boca Raton, pp 47–61
Erel Y, Morgan JJ (1992) The relationships between rock-derived lead and iron in natural waters. Geochim Cosmochim Acta 56:4157–4167
Farquhar GJ, Constable TW (1978) Leachate contaminant attenuation in soil, vol 1. Main Report, Project no 2123. University of Waterloo Research Institute, Waterloo, Ontario, Canada
Farrah H, Pickering WF (1978) The sorption of mercury species by clay minerals. Water Air Soil Pollut 9:23–31
Feick G, Horne RA, Yeapple D (1972) Release of mercury from contaminated freshwater sediments by the runoff of road deicing salt. Science 175:1142–1143
Fendorf SE, Sparks DL (1991) Chromium(III) induced solid phase transformation of δ-MnO_2 [abstract]. Program and Abstracts, 28th Annual Meeting of Clay Minerals Society, 5–10 Oct 1991, Houston, Texas. Lunar & Planetary Inst (LPI), Houston, LPI Contrib no 773, p 48
Figge DAH, Hetrick BAD, Wilson GWT (1995) Role of expanded clay and porous ceramic amendments on plant establishment in Mine spoils. Environ Pollut 88:161–165
Fimreite N, Reynolds LM (1973) Mercury contamination of fish in northwestern Ontario. J Wildl Manage 37:62–68
Fitzgerald WF (1979) Distribution of mercury in natural waters. In: Nriagu JO (ed) The biogeochemistry of mercury in the environment. Elsevier/North-Holland Biomedical Press, Amsterdam, pp 161–173
Flemming CA, Ferris FG, Beveridge TJ, Bailey GW (1990) Remobilization of toxic heavy metals adsorbed to bacterial wall–clay composites. Appl Environ Microbiol 56:3191–3203
Forbes EA, Posner AM, Quirk JP (1976) The specific adsorption of divalent Cd, Co, Cu, Pb, and Zn on goethite. J Soil Sci 27:154–166
Förstner U (1980) Cadmium in polluted sediments. In: Nriagu JO (ed) Cadmium in the environment, part 1 (ecological cycling). John Wiley, New York, pp 305–363
Förstner U (1991) Soil pollution phenomena – mobility of heavy metals in contaminated soil. In: Bolt GH, De Boodt MF, Hayes MHB, McBride MB, De Strooper EBA (eds) Interactions at the soil colloid–soil solution interface. Kluwer, Dordrecht, pp 543–582
Förstner U, Colombi C, Kistler R (1991) Dumping of wastes. In: Merian E (ed) Metals and their compounds in the environment. VCH, Weinheim, pp 333–355
Förstner U, Müller G (1973) Heavy metal accumulation in river sediments: a response to environmental pollution. Geoforum 14:53–61
Förstner U, Patchineelam SR (1976) Bindung und Mobilisation von Schwermetallen in fluviatilen Sedimenten. Chemiker Z 100:49–57

Förstner U, Patchineelam SR (1980) Chemical associations of heavy metals in polluted sediments from the lower Rhine River. In: Kavanaugh MC, Leckie JO (eds) Particulates in water. Advances in chemistry series, no 189. American Chemical Society, Washington, DC, pp 177–193
Förstner U, Wittmann GTW (1976) Metal accumulations in acidic waters from gold mines in South Africa. Geoforum 7:41–49
Fowler SW (1977) Trace elements in zooplankton particulate products. Nature 269:51–53
Francis AJ, Dodge CJ (1990) Anaerobic microbial remobilization of toxic metals coprecipitated with iron oxide. Environ Sci Technol 24:373–378
Fujiyoshi R, Eugene AS, Katayama M (1992) Behavior of radionuclides in the environment. I. Sorption of Zn(II) on clay minerals. Appl Radiat Isot (Int J Radiat Appl Instrum, part A) 43:1223–1226
Gadde RR, Laitinen HA (1974) Studies of heavy metal adsorption by hydrous iron and manganese oxides. Anal Chem 46:2022–2026
Gagnon C, Arnac M, Brindle J-R (1992) Sorption interactions between trace metals (Cd and Ni) and phenolic substances on suspended clay minerals. Water Res 26:1067–1072
García-Miragaya J, Page AL (1976) Influence of ionic strength and inorganic complex formation on the sorption of trace amounts of Cd by montmorillonite. Soil Sci Soc Am J 40:658–663
García-Miragaya J, Page AL (1977) Influence of exchangeable cation on the sorption of trace amounts of cadmium by montmorillonite. Soil Sci Soc Am J 41:718–721
Gavis J, Ferguson JF (1972) The cycling of mercury through the environment. Water Res 6:989–1008
Gibbs RJ (1977) Transport phases of transition metals in the Amazon and Yukon Rivers. Geol Soc Am Bull 88:829–843
Gobeil C, Cossa D (1993) Mercury in sediments and sediment pore water in the Laurentian Trough. Can J Fish Aquat Sci 50:1794–1800
Goldberg ED, Koide M, Bertine K, Hodge V, Stallard M, Martincic D, Mikac N, Branica M, Abaychi JK (1988) Marine geochemistry – 2: scavenging redux. Appl Geochem 3:561–571
Gomer JS, Yopps SW, Sandoval SP, Clark AE (1992) Copper exchange capacity of clays and their potential effect on in situ copper leaching. Report of Investigations 9396, Bureau of Mines, US Dept of the Interior, Pittsburgh, 10 pp
Goodman BA, Cheshire MV (1973) Electron paramagnetic resonance evidence that copper is complexed in humic acid by the nitrogen of porphyrin groups. Nature (New Biol) 244:158–159
Greenland DJ (1971) Interactions between humic and fulvic acids and clays. Soil Sci 111:34–41
Griffin RA, Frost RR, Au AK, Robinson GD, Shimp NF (1977) Attenuation of pollutants in municipal landfill leachate by clay minerals: part 2 – heavy metal adsorption. Environmental Geol Notes no 79, Illinois State Geological Survey, Urbana, 47 pp
Grim, RE (1968) Clay mineralogy, 2nd edn. McGraw-Hill, New York, 596 pp
de Groot AJ, Allersma E (1975) Field observations on the transport of heavy metals in sediments. In: Krenkel PA (ed) Heavy metals in the aquatic environment. Pergamon Press, Oxford, pp 85–101
de Groot AJ, de Goeij JJM, Zegers C (1971) Contents and behaviour of mercury as compared with other heavy metals in sediments from the rivers Rhine and Ems. Geol Mijnbouw 50:393–398

Gunn AM, Hunt DTE, Winnard DA (1989) The effect of heavy metal speciation in sediment on bioavailability to tubificid worms. Hydrobiologia 188/189: 487–496

Gupta GS, Singh AK, Tyagi BS, Prasad G, Singh VN (1992) Treatment of carpet and metallic effluents by china clay. Chem Tech Biotechnol 55:277–283

Haghiri F (1974) Plant uptake of cadmium as influenced by cation exchange capacity, organic matter, zinc, and soil temperature. J Environ Qual 3:180–183

Hahne HCH, Kroontje W (1973) Significance of pH and chloride concentration on behavior of heavy metal pollutants: mercury(II), zinc(II), and lead(II). J Environ Qual 2:444–450

Hamdy MK, Wheeler SR (1978) Inhibition of bacterial growth by mercury and the effects of protective agents. Bull Environ Contam Toxicol 20:378–386

Hamed J, Acar YB, Gale RJ (1983) Pb(II) removal from kaolinite by electrokinetics. J Geotech Engin 117:241–271

Hare L (1992) Aquatic insects and trace metals: bioavailability, bioaccumulation, and toxicity. Crit Rev Toxicol 22:327–369

Hare L, Saouter E, Campbell PGC, Tessier A, Ribeyre F, Boudou A (1991) Dynamics of cadmium, lead, and zinc exchange between nymphs of the burrowing mayfly *Hexagenia rigida* (Ephemeroptera) and the environment. Can J Fish Aquat Sci 48:39–47

Harmsen K (1977) Behaviour of heavy metals in soils. Centre for Agricultural Publishing and Documentation (Pudoc), Wageningen, 177 pp

Hayes MHB, Himes FL (1986) Nature and properties of humus-mineral complexes. In: Huang PM, Schnitzer M (eds) Interactions of soil minerals with natural organics and microbes. SSSA Special Publication no. 17, Soil Science Society of America, Madison, Wisconsin, pp 103–158

Hecky RE, Bodaly RA, Strange NE, Ramsey DJ, Anema C, Fudge RJP (1987) Mercury bioaccumulation in yellow perch in limnocorrals simulating the effects of reservoir formation. In: Technical Appendices to the Summary Report, Canada-Manitoba Agreement on the Study and Monitoring of Mercury in the Churchill River diversion, vol 2, chapt 7. Governments of Canada and Manitoba, Winnipeg, Manitoba, Canada

Hecky RE, Ramsey DJ, Bodaly RA, Strange NE (1991) Increased methylmercury contamination in fish in newly formed freshwater reservoirs. In: Suzuki T, Imura N, Clarkson TW (eds) Advances in mercury toxicology. Plenum Press, New York, pp 33–52

Hegeman WJM, van der Weijden CH, Zwolsman JJG (1992) Sorption of zinc on suspended particles along a salinity gradient: a laboratory study using illite and suspended matter from the river Rhine. Neth J Sea Res 28: 285–292

Helios-Ryicka E, Kyziol J (1991) Clays and clay minerals as the natural barries [sic] for heavy metals in pollution mechanisms – illustrated by Polish rivers and soils. Mitt Österr Geol Ges 83:163–176

Hem JD (1978) Redox processes at surfaces of manganese oxide and their effects on aqueous metal ions. Chem Geol 21:199–218

Hem JD, Lind CJ (1991) Coprecipitation mechanisms and products in manganese oxidation in the presence of cadmium. Geochim Cosmochim Acta 55:2435–2451

Hingston FJ (1981) A review of anion adsorption. In: Anderson MA, Rubin AJ (eds) Adsorption of inorganics at solid-liquid interfaces. Ann Arbor Science, Ann Arbor, pp 51–90

Hingston FJ, Atkinson RJ, Posner AM, Quirk JP (1968) Specific adsorptions of anions on goethite. Trans 9th Int Congr Soil Sci 1:669–678

Hingston FJ, Posner AM, Quirk JP (1971) Competitive adsorption of negatively charged ligands on oxide surfaces. Discussions Faraday Soc 52:334–342

Hingston FJ, Posner AM, Quirk, JP (1974) Anion adsorption by goethite and gibbsite. II. Desorption of anions from hydrous oxide surfaces. J Soil Sci 25:16–26

Hogg TJ, Stewart JWB, Bettany JR (1978) Influence of the chemical form of mercury on its adsorption and ability to leach through soils. J Environ Qual 7:440–445

Hoins U, Charlet L, Sticher H (1993) Ligand effect on the adsorption of heavy metals: the sulfate-cadmium-goethite case. Water Air Soil Pollut 68:241–255

Holmes CW, Slade EA, McLerran CJ (1974) Migration and redistribution of zinc and cadmium in marine estuarine system. Environ Sci Technol 8:255–259

Honeyman BD, Santschi PH (1992) The role of particles and colloids in the transport of radionuclides and trace metals in the oceans. In: Buffle J, van Leeuwen HP (eds) Environmental particles, vol 1. Lewis Publishers, Boca Raton, pp 379–423

Horowitz AJ, Lum KR, Garbarino JR, Hall GEM, Lemieux C, Demas CR (1996) Problems associated with using filtration to define dissolved trace element concentrations in natural water samples. Environ Sci Technol 30:954–963

Hsia TH, Lo SL, Lin CF (1992) Interaction of Cr(VI) with amorphous iron oxide: adsorption density and surface charge. Water Sci Technol 26:181–188

Huang PM, Liaw WK (1979) Adsorption of arsenite by lake sediments. Int Rev Gesamten Hydrobiol 64:263–271

Hunter RJ, James M (1992) Charge reversal of kaolinite by hydrolyzable metal ions: an electroacoustic study. Clays Clay Miner 40:644–649

Inoue Y, Munemori M (1979) Coprecipitation of mercury(II) with iron(III) hydroxide. Environ Sci Technol 13:443–445

Jackson TA (1971a) Evidence for selective adsorption and polymerization of the L-optical isomers of amino acids relative to the D-optical isomers on the edge faces of kaolinite. Experientia 27:242–243

Jackson TA (1971b) Preferential polymerization and adsorption of L-optical isomers of amino acids relative to D-optical isomers on kaolinite templates. Chem Geol 7:295–306

Jackson TA (1977) A relationship between crystallographic properties of illite and chemical properties of extractable organic matter in pre-Phanerozoic and Phanerozoic sediments. Clays Clay Minerals 25:187–195

Jackson TA (1978) The biogeochemistry of heavy metals in polluted lakes and streams at Flin Flon, Canada, and a proposed method for limiting heavy metal pollution of natural waters. Environ Geol 2:173–189

Jackson TA (1979) Relationships between the properties of heavy metals and their biogeochemical behaviour in lakes and river-lake systems. Proc Int Conf, London, September 1979, Management & Control of Heavy Metals in the Environment. CEP Consultants, Edinburgh, pp 457–460

Jackson TA (1986) Methyl mercury levels in a polluted prairie river-lake system: seasonal and site-specific variations, and the dominant influence of trophic conditions. Can J Fish Aquat Sci 43:1873–1887

Jackson TA (1987) Methylation, demethylation, and bio-accumulation of mercury in lakes and reservoirs of northern Manitoba, with particular reference to effects of environmental changes caused by the Churchill-Nelson River diversion. In: Technical Appendices to the Summary Report, Canada-Manitoba-Agreement on the Study and Monitoring of Mercury in the Churchill River Diversion, vol 2, chap 8. Governments of Canada and Manitoba, Winnipeg, Manitoba, Canada

Jackson TA (1988a) Accumulation of mercury by plankton and benthic invertebrates in riverine lakes of northern Manitoba (Canada): importance of regionally and seasonally varying environmental factors. Can J Fish Aquat Sci 45:1744–1757
Jackson TA (1988b) The mercury problem in recently formed reservoirs of northern Manitoba (Canada): effects of impoundment and other factors on the production of methyl mercury by microorganisms in sediments. Can J Fish Aquat Sci 45:97–121
Jackson TA (1989) The influence of clay minerals, oxides, and humic matter on the methylation and demethylation of mercury by micro-organisms in freshwater sediments. Appl Organometall Chem 3:1–30
Jackson TA (1991) Biological and environmental control of mercury accumulation by fish in lakes and reservoirs of northern Manitoba, Canada. Can J Fish Aquat Sci 48:2449–2470
Jackson TA (1993a) Comment on "Weathering, plants, and the long-term carbon cycle" by Robert A. Berner. Geochim Cosmochim Acta 57:2141–2144
Jackson TA (1993b) Effects of environmental factors and primary production on the distribution and methylation of mercury in a chain of highly eutrophic riverine lakes. Water Pollut Res J Can 28:177–216. Also see "Erratum," Water Pollut Res J Canada 28: unnumbered page following p 512 (1993)
Jackson TA (1993c) The influence of phytoplankton blooms and environmental variables on the methylation, demethylation, and bio-accumulation of mercury (Hg) in a chain of eutrophic mercury-polluted riverine lakes in Saskatchewan, Canada. In: Allan RJ, Nriagu JO (eds) Heavy Metals in the Environment (Proc Int Conf, Toronto, Sept 1993), vol 2. CEP Consultants, Edinburgh, pp 301–304
Jackson TA (1995) Effects of clay minerals, oxyhydroxides, and humic matter on microbial communities of soil, sediment, and water. In: Huang PM, Berthelin J, Bollag J-M, McGill WB, Page AL (eds) Environmental impact of soil component interactions. Lewis Publishers (CRC), Boca Raton, pp 165–200
Jackson TA, Bistricki T (1995) Selective scavenging of copper, zinc, lead, and arsenic by iron and manganese oxyhydroxide coatings on plankton in lakes polluted with mine and smelter wastes: results of energy dispersive X-ray micro-analysis. J Geochem Explor 52:97–125
Jackson TA, Hecky RE (1980) Depression of primary productivity by humic matter in lake and reservoir waters of the Boreal forest zone. Can J Fish Aquat Sci 37:2300–2317
Jackson TA, Kipphut G, Hesslein RH, Schindler DW (1980a) Experimental study of trace metal chemistry in soft-water lakes at different pH levels. Can J Fish Aquat Sci 37:387–402
Jackson TA, Parks JW, Rudd JWM (1980b) Discussion, with tentative conclusions and recommendations, based on results of the 1978–1979 study of mercury in the Wabigoon River system. In: Jackson TA (ed) Mercury pollution in the Wabigoon-English River system of northwestern Ontario, and possible remedial measures: a progress report. Governments of Canada and Ontario, Winnipeg, Manitoba, Canada
Jackson TA, Parks JW, Jones PD, Woychuk RN, Sutton JA, Hollinger JD (1982) Dissolved and suspended mercury species in the Wabigoon River (Ontario, Canada): seasonal and regional variations. Hydrobiologia 92:473–487
Jackson TA, Klaverkamp JF, Dutton MD (1993) Heavy metal speciation and its biological consequences in a group of lakes polluted by a smelter, Flin Flon, Manitoba, Canada. Appl Geochem (Suppl Issue) 2: pp 285–289

Jackson TA, Reynoldson TB, Zhou K, Nguyen N, Rosa F, Mudroch A (1995) Selective toxic effects of heavy metals on reproductive activities of the oligochaete *Tubifex tubifex* in lake sediments polluted with mine tailings. In: Wilken R-D, Förstner U, Knöchel A (eds) Heavy metals in the environment, vol 2. Proc 10th Int Conf on Heavy metals in the environment, Hamburg, September 1995. CEP Consultants, Edinburgh, pp 13–16

Jackson, TA, Voigt GK (1971) Biochemical weathering of calcium-bearing minerals by rhizosphere micro-organisms, and its influence on calcium accumulation in trees. Plant Soil 35:655–658

Jackson TA, Woychuk RN (1980a) Mercury speciation and distribution in a polluted river-lake system as related to the problem of lake restoration. In: Restoration of lakes and inland waters (Proc Int Symp on Inland Waters and Lake Restoration, 8–12 September 1980, Portland, Maine). Office of Water Regulations and Standards, US Environmental Protection Agency, Washington, DC, (EPA 440/5-81-010), pp 93–101

Jackson TA, Woychuk RN (1980b) The geochemistry and distribution of mercury in the Wabigoon River system. In: Jackson TA (ed) Mercury pollution in the Wabigoon-English River system of Northwestern Ontario, and possible remedial measures: a progress report. Governments of Canada and Ontario, Winnipeg, Manitoba, Canada

Jackson TA, Woychuk RN (1981) Methyl mercury formation and distribution in a polluted river-lake system: the effect of environmental variables, and implications for biological uptake and lake restoration [abstract]. Verh Int Ver Limnol 21:1114–1115

Jarvis SC (1981) Copper sorption by soils at low concentrations and relation to uptake by plants. J Soil Sci 32:257–269

Jenne EA (1968) Controls on Mn, Fe, Co, Ni, Cu, and Zn concentrations in soils and water: the significant role of hydrous Mn and Fe oxides. In: American Chemical Society (eds) Trace inorganics in water. Advances in chemistry series no 73. American Chemical Society, Washington, DC, pp 337–387

John MK (1972) Cadmium adsorption maxima of soils as measured by the Langmuir isotherm. Can J Soil Sci 52:343–350

Karimian N, Cox FR (1979) Molybdenum availability as predicted from selected soil chemical properties. Agron J 71:63–65

Karlsson S, Sandén P, Allard B (1987) Environmental impacts of an old mine tailings deposit - metal adsorption by particulate matter. Nord Hydrol 18:313–324

Kester DR, Andreae MO, Bernhard M, Branica M, Calmano W, Duinker JC, George SG, Lund W, Luoma SN, Tramier B, Velapoldi RA, Vestal ML (1986) Chemical species in marine and estuarine systems. In: Bernhard M, Brinckman FE, Sadler PJ (eds) The importance of chemical "speciation" in environmental processes. Springer, Berlin Heidelberg New York, pp 275–299 (Dahlem Konferenzen 1986)

Khalid RA (1980) Chemical mobility of cadmium in sediment-water systems. In: Nriagu JO (ed) Cadmium in the environment, Part 1 (ecological cycling). John Wiley, New York, pp 257–304

Kharkar DP, Turekian KK, Bertine KK (1968) Stream supply of dissolved silver, molybdenum, antimony, selenium, chromium, cobalt, rubidium and cesium to the oceans. Geochim Cosmochim Acta 32:285–298

Kidby DK, Murray AD, Jordan DC (1977) Toxicity of mercury at the molecular and whole-organism level: role of competitive binding. In: Drucker H, Wildung RE (eds) Biological implications of metals in the environment. Proc 15th Annu Hanford Life Sciences Symp, 29 Sept–1 Oct 1975, Richland, Washington. Technical Information Center, Energy Research and Development Administration, US Dept of Commerce, Springfield, Virginia, pp 69–76

Kieffer F (1991) Metals as essential trace elements for plants, animals, and humans. In: Merian E (ed) Metals and their compounds in the environment. VCH, Weinheim, pp 481–489

Kinniburgh DG (1983) The H^+/M^{2+} exchange stoichiometry of calcium and zinc adsorption by ferrihydrite. J Soil Sci 34:759–768

Kinniburgh DG, Jackson ML (1978) Adsorption of mercury(II) by iron hydrous oxide gel. Soil Sci Soc Am J 42:45–47

Kinniburgh DG, Jackson ML (1981) Cation adsorption by hydrous metal oxides and clay. In: Anderson MA, Rubin AJ (eds) Adsorption of inorganics at solid-liquid interfaces. Ann Arbor Science Publishers, Ann Arbor, pp 91–160

Kinniburgh DG, Jackson ML, Syers JK (1976) Adsorption of alkaline earth, transition, and heavy metal cations by hydrous oxide gels of iron and aluminum. Soil Sci Soc Am J 40:796–799

Kozar S, Bilinski H, Branica M, Schwuger MJ (1992) Adsorption of Cd(II) and Pb(II) on bentonite under estuarine and seawater conditions. Sci Total Environ 121: 203–216

Krishnamurthy S (1992) Humic substances and contaminant transport: a review. Northeast Geol 14:59–64

Kudo A, Hart JS (1974) Uptake of inorganic mercury by bed sediment. J Environ Qual 3:273–278

Kudo A, Miller DR, Townsend DR (1977) Mercury transport interacting with bed sediment movements. Prog Water Technol 9:923–935

Latterell JJ, Dowdy RH, Larson WE (1978) Correlation of extractable metals and metal uptake of snap beans grown on soil amended with sewage sludge. J Environ Qual 7:435–440

Laube V, Ramamoorthy S, Kushner DJ (1979) Mobilization and accumulation of sediment bound heavy metals by algae. Bull Environ Contam Toxicol 21:763–770

Leckie JO (1986) Adsorption and transformation of trace element species at sediment/water interfaces. In: Bernhard M, Brinckman FE, Sadler PJ (eds) The importance of chemical "speciation" in environmental processes. Springer, Berlin Heidelberg New York, pp 237–254

Leckie JO, Davis JA (1979) Aqueous environmental chemistry of copper. In: Nriagu JO (ed) Copper in the Environment, Part I (ecological cycling). John Wiley, New York, pp 89–121

Leckie JO, James RO (1974) Control mechanisms for trace metals in natural waters. In: Rubin AJ (ed), Aqueous-environmental chemistry of metals. Ann Arbor Science Publishers, Ann Arbor, pp 1–76

Leckie JO, Benjamin MM, Hayes K, Kaufman G, Altmann S (1980) Adsorption/coprecipitation of trace elements from water with iron oxyhydroxide. Final Report, EPRI CS-1513, Research Project 910-1. Electric Power Research Institute, Palo Alto, California

Lee GF (1975) Role of hydrous metal oxides in the transport of heavy metals in the environment. In: Krenkel PA (ed) Heavy metals in the aquatic environment. Pergamon Press, Oxford, pp 137–147

Leppard GG (1992) Evaluation of electron microscope techniques for the description of aquatic colloids. In: Buffle J, van Leeuwen HP (eds) Environmental particles, vol 1. Lewis, Boca Raton, pp 231–289
Leppard GG, Buffle J, De Vitre RR, Perret D (1988) The ultrastructure and physical characteristics of a distinctive colloidal iron particulate isolated from a small eutrophic lake. Arch Hydrobiol 113:405–424. (Also see "Corrigenda," Arch Hydrobiol 115:160 (1989)
Leppard GG, de Vitre RR, Perret D, Buffle J (1989) Colloidal iron oxyhydroxy-phosphate: the sizing and morphology of an amorphous species in relation to partitioning phenomena. Sci Total Environ 87/88:345–354
Leppard GG, Burnison BK, Buffle J (1990) Transmission electron microscopy of the natural organic matter of surface waters. Anal Chim Acta 232:107–121
Lewis AG, Whitfield PH, Ramnarine A (1972) Some particulate and soluble agents affecting the relationship between metal toxicity and organism survival in the calanoid copepod *Euchaeta japonica*. Mar Biol 17:215–221
Liang L, McNabb JA, Paulk JM, Gu B, McCarthy JF (1993) Kinetics of Fe(II) oxygenation at low partial pressure of oxygen in the presence of natural organic matter. Environ Sci Technol 27:1864–1870
Lide DR (ed) (1992) CRC handbook of physics and chemistry, 73rd edn. CRC Press, Boca Raton
Lind CJ, Anderson LD (1992) Trace metal scavenging by precipitating Mn and Fe oxides. In: Kharaka YK, Maest AS (eds) Water-rock interaction, vol 1. Low temperature environments. AA Balkema, Rotterdam, pp 397–402
Lindqvist O, Johansson K, Aastrup M, Andersson A, Bringmark L, Hovsenius G, Håkanson L, Iverfeldt Å, Meili M, Timm B (1991) Mercury in the Swedish environment. Water Air Soil Pollut 55:1–261
Lindsay WL, Norvell WA (1978) Development of a DTPA soil test for zinc, iron, manganese, and copper. Soil Sci Soc Am J 42:421–428
Lion LW, Altmann RS, Leckie JO (1982) Trace-element adsorption characteristics of estuarine particulate matter: evaluation of contributions of Fe/Mn oxide and organic surface coatings. Environ Sci Technol 16:660–666
Livens FR, Baxter MS, Allen SE (1986) Physico-chemical associations of plutonium in Cumbrian soils. In: Bulman RA, Cooper JR (eds) Speciation of fission and activation products in the environment. Elsevier Applied Science, London, pp 143–150
Lockhart WL, Wagemann R, Tracey B, Sutherland D, Thomas DJ (1992) Presence and implications of chemical contaminants in the freshwaters of the Canadian Arctic. Sci Total Environ 122:165–243
Lockwood RA, Chen KY (1973) Adsorption of Hg(II) by hydrous manganese oxides. Environ Sci Technol 7:1028–1034
Loganathan P, Burau RG, Fuerstenau DW (1977) Influence of pH on the sorption of Co^{2+}, Zn^{2+}, and Ca^{2+} by a hydrous manganese oxide. Soil Sci Soc Am J 41:57–62
Lorenz H (1979) Binding forms of toxic heavy metals, mechanisms of entrance of heavy metals into the food chain, and possible measures to reduce levels in foodstuff. Ecotoxicol Environ Safety 3:47–58
Luoma SN, Jenne EA (1977) The availability of sediment-bound cobalt, silver, and zinc to a deposit-feeding clam. In: Drucker H, Wildung RE (eds) Biological implications of metals in the environment. Proc 15th Annu Hanford Life Sciences Symp, 29 Sept – 1 Oct 1975, Richland, Washington. Technical Information Center, Energy Research and Development Administration, US Dept of Commerce, Springfield, Virginia, pp 213–230

Madrid L, Diaz-Barrientos E, Contreras MC (1991) Relationships between zinc and phosphate adsorption on montmorillonite and an iron oxyhydroxide. Australian J Soil Res 29:239–247

Maes A, Cremers A (1986) Radionuclide sorption in soils and sediments: oxide-organic matter competition. In: Bulman RA, Cooper JR (eds) Speciation of fission and activation products in the environment. Elsevier Applied Science Publishers, London, pp 93–100

Majone M, Papini MP, Rolle E (1993) Clay adsorption of lead from landfill leachate. Environ Technol 14:629–638

Malla PB, Ravindranathan P, Komarneni S, Roy R (1991) Intercalation of copper metal clusters in montmorillonite. Nature 351:555–557

Manceau A, Charlet L, Boisset MC, Didier B, Spadini L (1992) Sorption and speciation of heavy metals on hydrous Fe and Mn oxides. From microscopic to macroscopic. Appl Clay Sci 7:201–223

Marshall CE (1964) The physical chemistry and mineralogy of soils, vol 1. Soil materials. John Wiley, New York, 388 pp

Marsi (1992) Illite, humic fractions, and humic-illite complexes: their metal complexation mechanisms. PhD Thesis, University of Kentucky, Lexington

Martin MH, Coughtrey PJ (1982) Biological monitoring of heavy metal pollution. Applied Science Publishers, London, 475 pp

Mayer LM, Schick LL (1981) Removal of hexavalent chromium from estuarine waters by model substrates and natural sediments. Environ Sci Technol 15:1482–1484

McBride MB (1976) Exchange and hydration properties of Cu^{2+} on mixed-ion Na^{+}–Cu^{2+} smectites. Soil Sci Am J 40:452–456

McBride MB (1978) Copper(II) interactions with kaolinite: factors controlling adsorption. Clays Clay Miner 26:101–106

McBride MB (1980) Interpretation of the variability of selectivity coefficients for exchange between ions of unequal charge on smectites. Clays Clay Miner 28: 255–261

McBride MB (1991a) Comments on "Adsorption of heavy metals by silicon and aluminum oxide surfaces on clay minerals." Soil Sci Soc Am J 55:1508–1509

McBride MB (1991b) Processes of heavy and transition metal sorption by soil minerals. In: Bolt GH, De Boodt MF, Hayes MHB, McBride MB, De Strooper EBA (eds) Interactions at the soil colloid-soil solution interface. Kluwer, Dordrecht, pp 149–175

McBride MB, Mortland MM (1974) Copper(II) interactions with montmorillonite: evidence from physical methods. Soil Sci Soc Am Proc 38:408–415

McKenzie RM (1980) The adsorption of lead and other heavy metals on oxides of manganese and iron. Australian J Soil Res 18:61–73

McKnight DM, Wershaw RL, Bencala KE, Zellweger GW, Feder GL (1992) Humic substances and trace metals associated with Fe and Al oxides deposited in an acidic mountain stream. Sci Total Environ 117/118:485–498

McLaren RG, Swift RS, Williams JG (1981) The adsorption of copper by soil materials at low equilibrium solution concentrations. J Soil Sci 32:247–256

McLaren RG, Lawson DM, Swift RS (1986) Sorption and desorption of cobalt by soils and soil components. J Soil Sci 37:413–426

Merian E (ed) (1991) Metals and their compounds in the environment. VCH, Weinheim, 1438 pp

Metwally AI, Mashhady AS, Falatah AM, Reda M (1993) Effect of pH on zinc adsorption and solubility in suspensions of different clays and soils. Z Pflanzenernähr Bodenkd 156:131–135

Modak DP, Singh KP, Chandra H, Ray PK (1992) Mobile and bound forms of trace metals in sediments of the lower Ganges. Water Res 26:1541–1548

Moore JN, Walker JR, Hayes TH (1990) Reaction scheme for the oxidation of As(III) to As(V) by birnessite. Clays Clay Miner 38:549–555

Morel FMM, Gschwend PM (1987) The role of colloids in the partitioning of solutes in natural waters. In: Stumm W (ed) Aquatic surface chemistry. John Wiley, New York, pp 405–422

Morgan JJ, Stumm W (1991) Chemical processes in the environment, relevance of chemical speciation. In: Merian E (ed) Metals and their compounds in the environment. VCH, Weinheim, pp 67–103

Morris C (ed) (1992) Academic Press dictionary of science and technology. Academic Press, San Diego, 2432 pp

Mosser C, Mosser A, Romeo M, Petit S, Decarreau A (1992) Natural and synthetic copper phyllosilicates studied by XPS. Clays Clay Miner 40:593–599

Mott HV, Singh S, Kondapally VR (1993) Factors affecting radium removal using mixed iron-manganese oxides. J Am Water Works Assoc 85:114–121

Muir DCG, Wagemann R., Hargrave BT, Thomas DJ, Peakall DB, Norstrom RJ (1992) Arctic marine ecosystem contamination. Sci Total Environ 122:75–134

Murozumi M, Chow TJ, Patterson C (1969) Chemical concentrations of pollutant lead aerosols, terrestrial dusts and sea salts in Greenland and Antarctic snow strata. Geochim Cosmochim Acta 33:1247–1294

Narkis N, Rebhun M, Sperber H (1968) Flocculation of clay suspensions in the presence of humic and fulvic acids. Isr J Chem 6:295–305

Newton DW, Ellis R Jr, Paulsen GM (1976) Effect of pH and complex formation on mercury(II) adsorption by bentonite. J Environ Qual 5:251–254

Nielsen JD (1990) Specific zinc adsorption as related to the composition and properties of clay and silt in some Danish soils. Acta Agric Scand 40:3–9.

Nriagu JO (1989) A global assessment of natural sources of atmospheric trace metals. Nature 338:47–49

Nriagu JO, Pacyna JM (1988) Quantitative assessment of worldwide contamination of air, water and soils by trace metals. Nature 333:134–139

Ohnesorge FK, Wilhelm M (1991) Zinc. In: Merian E (ed) Metals and their compounds in the environment. VCH, Weinheim, pp 1309–1342

Olson BH, Panigrahi AK (1991) Bacteria, fungi, and blue-green algae. In: Merian E (ed) Metals and their compounds in the environment. VCH, Weinheim, pp 419–448

Ongley ED, Blachford DP (1982) Application of continuous flow centrifugation to contaminant analysis of suspended sediment in fluvial systems. Environ Technol Lett 3:219–228

Osaki S, Miyoshi T, Sugihara S, Takashima Y (1990) Adsorption of Fe(III), Co(II) and Zn(II) onto particulates in fresh waters on the basis of the surface complexation model. II. Stabilities of metal species dissolved in fresh waters. Sci Total Environ 99:115–123

Oscarson DW, Huang PM, Defosse C, Herbillon A (1981a) Oxidative power of Mn(IV) and Fe(III) oxides with respect to As(III) in terrestrial and aquatic environments. Nature 291:50–51

Oscarson DW, Huang PM, Liaw WK (1981b) Role of manganese in the oxidation of arsenite by freshwater lake sediments. Clays Clay Miner 29:219–225
Oscarson DW, Huang PM, Liaw WK, Hammer UT (1983) Kinetics of oxidation of arsenite by various manganese dioxides. Soil Sci Soc Am J 47:644–648
Oscarson DW, Liaw WK, Huang PM (1981c) The kinetics and components involved in the oxidation of arsenite by freshwater lake sediments. Verh Int Ver Limnol 21:181–186
Painter S, Cameron EM, Allan R, Rouse J (1994) Reconnaissance geochemistry and its environmental relevance. J Geochem Explor 51:213–246
Parks JW, Craig PJ, Neary BP, Ozburn G, Romani D (1992) Biomonitoring in the mercury-contaminated Wabigoon-English-Winnipeg River (Canada) system: selecting the best available bioindicator. Appl Organometall Chem 5:487–495
Parks JW, Curry C, Romani D, Russell DD (1991) Young northern pike, yellow perch and crayfish as bioindicators in a mercury contaminated watercourse. Environ Monit Assess 16:39–73
Parks JW, Hamilton AL (1987) Accelerating recovery of the mercury-contaminated Wabigoon/English River system. Hydrobiologia 149:159–188
Patrick WH, Gambrell RP, Khalid RA (1977) Physicochemical factors regulating solubility and bioavailability of toxic heavy metals in contaminated dredged sediment. J Environ Sci Health A12:475–492
Pauling L (1960) The nature of the chemical bond. Cornell University Press, Ithaca, 644 pp
Payne K, Pickering WF (1975) Influence of clay-solute interactions on aqueous copper ion levels. Water Air Soil Pollut 5:63–69
Perhac RM (1972) Distribution of Cd, Co, Cu, Fe, Mn, Ni, Pb and Zn in dissolved and particulate solids from two streams in Tennessee. J Hydrol 15:177–186
Peters RW, Shem L (1992) Adsorption/desorption characteristics of lead on various types of soil. Environ Progr 11:234–240
Petersen W, Wallmann K, Schröer, Schroeder F (1993) Studies on the adsorption of cadmium on hydrous iron(III) oxides in oxic sediments. Anal Chim Acta 273:323–327
Phillips CSG, Williams RJP (1965) Inorganic chemistry, vol 1. Oxford University Press, Oxford, 685 pp
Pickering WF (1979) Copper retention by soil/sediment components. In: Nriagu JO (ed) Copper in the environment. I. Ecological cycling. John Wiley, New York, pp 217–253
Pickering WF (1980a) Cadmium retention by clays and other soil or sediment components. In: Nriagu JO (ed) Cadmium in the environment, part I. Ecological cycling. John Wiley, New York, pp 365–397
Pickering WF (1980b) Zinc interaction with soil and sediment components. In: Nriagu JO (ed) Zinc in the environment, part I. Ecological cycling. John Wiley, New York, pp 71–112
Puls RW, Powell RM, Clark D, Eldred CJ (1991) Effects of pH, solid/solution ratio, ionic strength, and organic acids on Pb and Cd sorption on kaolinite. Water Air Soil Pollut 57–58:423–430
Rae JE, Aston SR (1982) The role of suspended solids in the estuarine geochemistry of mercury. Water Res 16:649–654
Rashid MA, Buckley DE, Robertson KR (1972) Interactions of a marine humic acid with clay minerals and a natural sediment. Geoderma 8:11–27

Rasmussen PE (1994) Current methods of estimating atmospheric mercury fluxes in remote areas. Environ Sci Technol 28:2233–2241
Reimers RS, Krenkel PA (1974) Kinetics of mercury adsorption and desorption in sediments. Water Pollut Control Fed 46:352–365
Richman LA, Wren CD, Stokes PM (1988) Facts and fallacies concerning mercury uptake by fish in acid stressed lakes. Water Air Soil Pollut 37:465–473
Richter RO, Theis TL (1980) Nickel speciation in a soil/water system. In: Nriagu JO (ed) Nickel in the environment. John Wiley, New York, pp 189–202
Rickard DT, Nriagu JO (1978) Aqueous environmental chemistry of lead. In: Nriagu JO (ed) The biogeochemistry of lead in the environment. Elsevier/North-Holland Biomedical Press, Amsterdam, pp 219–284
Rogers JS, Huang PM, Hammer UT, Liaw WK (1984) Dynamics of desorption of mercury adsorbed on poorly crystalline oxides of manganese, iron, aluminium, and silicon. Verh Int Ver Limnol 22:283–288
Roy WR, Krapac IG, Steele JD (1993) Sorption of cadmium and lead by clays from municipal incinerator ash-water suspensions. J Environ Qual 22:537–543
Rudd JWM, Turner MA (1983) The English–Wabigoon River system. II. Suppression of mercury and selenium bioaccumulation by suspended and bottom sediments. Can J Fish Aquat Sci 40:2218–2227
Sabet SA, Wassif MA, El Kadi MA (1975) Availability of Zn and Cu in highly calcareous soils I. Effect of pure clay minerals and soil clay separates. Egypt J Soil Sci (Spec Issue): 351–361
Sakurai K, Huang PM (1995) Cadmium adsorption on the hydroxyaluminum–montmorillonite complex as influenced by oxalate. In: Huang PM, Berthelin J, Bollag J-M, McGill WB, Page AL (eds) Environmental impact of soil component interactions. Lewis (CRC), Boca Raton, pp 39–46
Salomons W, Baccini P (1986) Chemical species and metal transport in lakes. In: Bernhard M, Brinckman FE, Sadler PJ (eds) The importance of chemical "speciation" in environmental processes. Dahlem Konferenzen 1986, Springer, Berlin Heidelberg New York, pp 193–216
Salomons W, de Groot AJ (1978) Pollution history of trace metals in sediments, as affected by the Rhine River. In: Krumbein WE (ed) Environmental biogeochemistry and geomicrobiology, vol 1 (The aquatic environment). Ann Arbor Science Publishers, Ann Arbor, pp 149–162
Sato Y (1989) The rate of oxidation of ferrous iron in seawater and the partition of elements between iron oxides and seawater. J Oceanogr Soc Japan 45:270–278
Scheinberg H (1991) Copper. In: Merian E (ed) Metals and their compounds in the environment. VCH, Weinheim, pp 893–907
Schindler PW (1981) Surface complexes at oxide–water interfaces. In: Anderson MA, Rubin AJ (eds) Adsorption of inorganics at solid-liquid interfaces. Ann Arbor Science Publishers, Ann Arbor, pp 1–49
Schindler PW, Sposito G (1991) Surface complexation at (hydr)oxide surfaces. In: Bolt GH, De Boodt MF, Hayes MHB, McBride MB, De Strooper EBA (eds) Interactions at the soil colloid–soil solution interface. Kluwer, Dordrecht, pp 115–145
Schindler PW, Stumm W (1987) The surface chemistry of oxides, hydroxides, and oxide minerals. In: Stumm W (ed) Aquatic surface chemistry. John Wiley, New York, pp 83–110
Schmitt HW, Sticher H (1991) Heavy metal compounds in the soil. In: Merian E (ed) Metals and their compounds in the environment. VHC, Weinheim, pp 311–331

Schnitzer M, Khan SU (1972) Humic substances in the environment. Marcel Dekker, New York, 327 pp

Schulthess CP, Huang CP (1990) Adsorption of heavy metals by silicon and aluminum oxide surfaces on clay minerals. Soil Sci Soc Am J 54:679–688

Schulthess CP, Huang CP (1991) Reply to "Comments on 'Adsorption of heavy metals by silicon and aluminum oxide surfaces on clay minerals.'" Soil Sci Soc Am J 55:1509–1510

Schuster E (1991) The behavior of mercury in the soil with special emphasis on complexation and adsorption processes – a review of the literature. Water Air Soil Pollut 56:667–680

Schütte KH (1964) The biology of the trace elements. JB Lippincott, Philadelphia, 228 pp

Schwertmann U, Kodama H, Fischer WR (1986) Mutual interactions between organics and iron oxides. In: Huang PM, Schnitzer M (eds) Interactions of soil minerals with natural organics and microbes. SSSA Spec Pub 17. Soil Science Society of America, Madison, Wisconsion, pp 223–250

Sharma YC, Prasad G, Rupainwar DC (1991) Removal of Ni(II) from aqueous solutions by sorption. Int J Environ Stud 37:183–191

Shuman LM (1975) The effect of soil properties on zinc adsorption by soils. Soil Sci Soc Am Proc 39:454–458

Shuman LM (1977) Adsorption of Zn by Fe and Al hydrous oxides as influenced by aging and pH. Soil Sci Soc Am J 41:703–706

Siegel BZ, Siegel SM (1979) Biological indicators of atmospheric mercury. In: Nriagu JO (ed), The biogeochemistry of mercury in the environment. Elsevier/North Holland Biomedical Press, Amsterdam, pp 131–159

Sikalidis CA, Alexiades C, Misaelides P (1989) Adsorption of uranium and thorium from aqueous solutions by the clay minerals montmorillonite and vermiculite. Toxicol Environ Chem 20–21:175–180

Sillén LG, Martell AE (1964) Stability constants of metal-ion complexes. Spec Publ no 17. The Chemical Society, London

Sillén LG, Martell AE (1971) Stability constants of metal-ion complexes, Suppl 1. Spec Publ no 25. The Chemical Society, London

Sinicrope TL, Langis R, Gersberg RM, Busnardo MJ, Zedler JB (1992) Metal removal by wetland mesocosms subjected to different hydroperiods. Ecol Engin 1: 309–322

Slemr F, Langer E (1992) Increase in global atmospheric concentrations of mercury inferred from measurements over the Atlantic Ocean. Nature 355:434–437

Snodgrass WJ (1980) Distribution and behaviour of nickel in the aquatic environment. In: Nriagu JO (ed) Nickel in the environment. John Wiley, New York, pp 203–274

Spencer DW, Brewer PG, Sachs PL (1972) Aspects of the distribution and trace element composition of suspended matter in the Black Sea. Geochim Cosmochim Acta 36:71–86

Srivastava SK, Tyagi R, Pant N, Pal N (1989) Studies on the removal of some toxic metal ions. Part II (Removal of lead and cadmium by montmorillonite and kaolinite). Environ Technol Lett 10:275–282

Stadler M, Schindler PW (1993a) Modeling of H^+ and Cu^{2+} adsorption on calcium-montmorillonite. Clays Clay Miner 41:288–296

Stadler M, Schindler PW (1993b) The effect of dissolved ligands upon the sorption of Cu(II) by Ca-montmorillonite. Clays Clay Miner 41:680–692

Stiff AC, Fernando Q, Zeitlin H (1982) Electron microprobe analysis of Pacific Ocean ferromanganese nodules. Mar Min 3:271–284

Stockmeyer M, Kruse K (1991) Adsorption of Zn and Ni ions and phenol and diethylketones by bentonites of different organophilicities. Clays Clay Miner 26: 431–434

Stoeppler M (1991) Cadmium. In: Merian E (ed) Metals and their compounds in the environment. VCH Weinheim, pp 803–851

Stotzky G (1986) Influence of soil mineral colloids on metabolic processes, growth, adhesion, and ecology of microbes and viruses. In: Huang PM, Schnitzer M (eds) Interactions of soil minerals with natural organics and microbes. SSSA Spec Publ 17, Soil Science Society of America, Madison, Wisconsin, pp 305–428

Street JJ, Lindsay WL, Sabey BR (1977) Solubility and plant uptake of cadmium in soils amended with cadmium and sewage sludge. J Environ Qual 6:72–77

Stumm W, Hohl H, Dalang F (1976) Interaction of metal ions with hydrous oxide surfaces. Croat Chem Acta 48:491–504

Sun Kou MR, Menioroz S, Fierro JLG, Rodriguez-Ramos I, Palacios JM, Guerrero-Ruiz, De Andres AM (1992) Naturally-occurring silicates as carriers for copper catalysts used in methanol conversion. Clays Clay Miner 40:167–174

Swift RS, McLaren RG (1991) Micronutrient adsorption by soils and soil colloids. In: Bolt GH, De Boodt MF, Hayes MHB, McBride MB, De Strooper EBA (eds) Interactions at the soil colloid–soil solution interface. Kluwer, Dordrecht, pp 257–292

Tam S-C, Chow A, Hadley D (1995) Effects of organic component on the immobilization of selenium on iron oxyhydroxide. Sci Total Environ 164:1–7

Tessier A, Campbell PGC, Auclair JC, Bisson M (1984) Relationships between the partitioning of trace metals in sediments and their accumulation in the tissues of the freshwater mollusc *Elliptio complanata* in a mining area. Can J Fish Aquat Sci 41:1463–1472

Tessier A, Carignan R, Dubreuil B, Rapin F (1989) Partitioning of zinc between the water column and the oxic sediments in lakes. Geochim Cosmochim Acta 53: 1511–1522

Tessier A, Couillard Y, Campbell PGC, Auclair JC (1993) Modeling Cd partitioning in oxic lake sediments and Cd concentrations in the freshwater bivalve *Anodonta grandis*. Limnol Oceanogr 38:1–17

Tessier A, Fortin D, Belzile N, DeVitre RR, Leppard GG (1996) Metal sorption to diagenetic iron and manganese oxyhydroxides and associated organic matter: narrowing the gap between field and laboratory measurements. Geochim Cosmochim Acta 60:387–404

Thanabalasingam P, Pickering WF (1990) Specific sorption of antimony (III) by the hydrous oxides of Mn, Fe, and Al. Water Air Soil Pollut 49:175–185

Theng BKG (1974) The chemistry of clay-organic reactions. John Wiley, New York, 343 pp

Thomas RL (1972) The distribution of mercury in the sediments of Lake Ontario. Can J Earth Sci 9:636–651

Thornton I (1979) Copper in soils and sediments. In: Nriagu JO (ed) Copper in the environment, part I. Ecological cycling. John Wiley, New York, pp 171–216

Tiller KG, Honeysett JL, de Vries MPC (1972) Soil zinc and its uptake by plants. Australian J Soil Res 10:165–182

Tipping E (1981) The adsorption of aquatic humic substances by iron oxides. Geochim Cosmochim Acta 45:191–199
Traina SJ, Doner HE (1985) Copper-manganese(II) exchange on a chemically reduced birnessite. Soil Sci Soc Am J 49:307–313
Trolard F, Bourrie G, Jeanroy E, Herbillon AJ, Martin H (1995) Trace metals in natural iron oxides from laterites: a study using selective kinetic extraction. Geochim Cosmochim Acta 59:1285–1297
Turekian KK (1977) The fate of metals in the ocean. Geochim Cosmochim Acta 41: 1139–1144
Turner A, Tyler AO (1997) Modelling adsorption and desorption processes in estuaries, chap 3 In: Jickells T, Rae JE (eds) Biogeochemistry of intertidal sediments. Cambridge University Press, Cambridge
Urrutia MM, Beveridge TJ (1993) Remobilization of heavy metals retained as oxyhydroxides or silicates by *Bacillus subtilis* cells. Appl Environ Microbiol 59: 4323–4329
van der Weijden CH (1990) Behaviour of heavy metals upon transition from riverine to marine environments [abstract]. Program and Abstr, VM Goldschmidt Conf, 2–4 May 1990, Baltimore, p 88
van Driel W, Nijssen JPJ (1988) Development of dredged material disposal sites: implications for soil, flora and food quality. In: Salomons W, Förstner U (eds) Chemistry and biology of solid waste. Springer, Berlin Heidelberg New York, pp 101–126
van Olphen H (1963) Clay colloid chemistry. Interscience (John Wiley), New York, 301 pp
Wagemann R, Stewart REA (1994) Concentrations of heavy metals and selenium in tissues and some foods of walrus (*Odobenus rosmarus rosmarus*) from the eastern Canadian Arctic and sub-arctic, and associations between metals, age, and gender. Can J Fish Aquat Sci 51:426–436
Wagner J-F (1991) Retention of heavy metals from blast-furnace dedusting sludges by a clayey subsoil. Water Air Soil Pollut 57–58:351–357
Wagner J-F, Czurda KA (1991) Radionuklidsorption an tertiären Tonen. Mitt Österr Geol Ges 83:215–227
Wang JS, Huang PM, Hammer UT, Liaw WK (1985) Influence of selected cation and anion species on the adsorption of mercury(II) by montmorillonite. Appl Clay Sci 1:125–132
Wang JS, Huang PM, Hammer UT, Liaw WK (1988) Influence of chloride/mercury molar ratio and pH on the adsorption of mercury by poorly crystalline oxides of Al, Fe, Mn, and Si. Verh Int Ver Limnol 23:1594–1600
Wang JS, Huang PM, Hammer UT, Liaw WK (1989) Role of dissolved oxygen in the desorption of mercury from freshwater sediment. In: Nriagu JO (ed) Aquatic toxicology and water quality management. John Wiley, New York, pp 153–159.
Wang JS, Huang PM, Liaw WK, Hammer UT (1991) Kinetics of the desorption of mercury from selected freshwater sediments as influenced by chloride. Water Air Soil Pollut 56:553–542
Wang MC, Huang PM (1987) Polycondensation of pyrogallol and glycine and the associated reactions as catalysed by birnessite. Sci Total Environ 62:435–442
Ward NE, Bassett RL (1991) Adsorption of hexavalent chromium on kaolinite and illite. Proc Univ of Missouri's 22nd Annu Conf, 23–26 May 1988, St Louis. Trace Substances Environ Health 24:193–206

Warith MA, Yong RN (1991) Landfill leachate attenuation by clay soil. Hazardous Waste Hazardous Mater 8:127–141
Warren LA, Zimmerman, AP (1994) The importance of surface area in metal sorption by oxides and organic matter in a heterogeneous natural sediment. Appl Geochem 9:245–254
Waslenchuk DG (1975) Mercury in fluvial bed sediments subsequent to contamination. Environ Geol 1:131–136
Weast RC, Astle MJ (eds) (1982) CRC handbook of chemistry and physics. CRC Press, Boca Raton
Weatherley AH, Lake PS, Rogers SC (1980) Zinc pollution and the ecology of the freshwater environment. In: Nriagu JO (ed) Zinc in the environment, part I. Ecological cycling. John Wiley, New York, pp 337–418
Weiss HV, Koide M, Goldberg ED (1971) Mercury in a Greenland ice sheet: evidence of recent input by man. Science 174:692–694
Westall JC (1987) Adsorption mechanisms in aquatic surface chemistry. In: Stumm W (ed) Aquatic surface chemistry. John Wiley, New York, pp 3–32
Whitehouse UG, Jeffrey LM, Debbrecht JD (1960) Differential settling tendencies of clay minerals in saline waters. Clays Clay Miner 5:1–79
Wilke B-M (1991) Einfluss verschiedener Bodeneigenschaften auf die mikrobielle Toxizität von Blei und Cadmium. Z Pflanzenernähr Bodenkd 154:417–424
Wilkins BT, Green N, Stewart SP, Major RO (1986) Factors that affect the association of radionuclides with soil phases. In: Bulman RA, Cooper JR (eds) Speciation of fission and activation products in the environment. Elsevier Applied Science, London, pp 101–113
Windom HL, Kendall DR (1979) Accumulation and biotransformation of mercury in coastal and marine biota. In: Nriagu JO (ed) The biogeochemistry of mercury in the environment. Elsevier/North-Holland Biomedical Press, Amsterdam, pp 303–323
Winfrey MR, Rudd JWM (1990) Environmental factors affecting the formation of methylmercury in low pH lakes. Environ Toxicol Chem 9:853–869
Xing GX, Xu LY, Hou WH (1995) Role of amorphous Fe oxides in controlling retention of heavy metal elements in soils. In: Huang PM, Berthelin J, Bollag J-M, McGill WB, Page AL (eds) Environmental impact of soil component interactions. Lewis (CRC), Boca Raton, pp 63–74
Xu H, Allard B (1991) Effects of a fulvic acid on the speciation and mobility of mercury in aqueous solutions. Water Air Soil Pollut 56:709–717
Young LB, Harvey HH (1992) The relative importance of manganese and iron oxides and organic matter in the sorption of trace metals by surficial lake sediments. Geochim Cosmochim Acta 56:1175–1186
Young DR, Jan T-K, Hershelman GP (1980) Cycling of zinc in the nearshore marine environment. In: Nriagu JO (ed) Zinc in the environment, part I. Ecological cycling. John Wiley, New York, pp 297–335
Zachara JM, Cowan CE, Smith SC (1991) Multiple-site modeling of metal ion adsorption on specimen and soil layer silicates [Abstr]. Program and Abstr, 28th Annu Meeting, Clay Minerals Society, 5–10 Oct 1991, Houston. Lunar & Planetary Institute (LPI), Houston, p 180 (LPI Contrib no 773)
Zachara JM, Smith SC, Resch CT, Cowan CE (1992) Cadmium sorption to soil separates containing layer silicates and iron and aluminum oxides. Soil Sci Soc Am J 56:1074–1084

Zaitseva LV, Varentsov IM (1990) Influence of major ions of seawater (chloride systems) on Cu(II) sorption by manganese oxyhydroxides: use of experimental data in a model of polymetallic ore formation in Recent basins. Chem Erde 50: 255–268

Zechmeister HG (1995) Correlation between altitude and heavy metal deposition in the Alps. Environ Pollut 89:73–80

Zhuang Y, Allen HE, Gu G (1994) Effect of aeration of sediment on cadmium binding. Environ Toxicol Chem 13:717–724

Long-range atmospheric transport of Hg, which was mentioned briefly in Section 5.1, is treated more fully in the following recently published review paper:

Jackson TA (1997) Long-range atmospheric transport of mercury to ecosystems, and the importance of anthropogenic emissions – a critical review and evaluation of the published evidence. Environ Rev 5:99–120. Also see "Erratum", Environ Rev (in press)

For a comprehensive review of the biogeochemistry and biological effects of Hg in aquatic environments, with detailed treatment of subjects such as the elemental properties, reactions, complexing, sorption, desorption, and bioavailability of Hg, see the following:

Jackson TA, Mercury in aquatic ecosystems. Chapt. 5 in: Langston WJ, Bebianno MJ (eds) Metal metabolism in aquatic environments. Chapman & Hall, London (in press)

Chapter 6

Clay Barriers in Landfills

J. Arch

6.1 Introduction

This chapter presents a practical review of the use of clays as environmental barriers around waste disposal (landfill) sites. Clay barriers are generally used as liners and capping materials for landfill sites. In each case they isolate potentially polluting wastes from the surrounding environment such that the environmental impacts attributable to a landfill are minimised. This review initially discusses the hazards associated with waste disposal sites which necessitate the containment of wastes and describes why natural clays are so often used as barriers around landfills (Sect. 6.2). The physical and chemical properties of clays which render them suitable for use in landfills are then discussed (Sects. 6.3 and 6.4). A description of the practicalities involved in clay liner construction is then presented, including investigation and testing required prior to construction, and details of appropriate construction methods (Sect. 6.5). Construction quality control (CQC) and construction quality assurance (CQA) procedures to be followed during construction are then considered in detail in Section 6.6. Finally, some of the 'aftercare' problems which can occur on landfill sites following their development are described (6.6.5), with particular emphasis on those concerning any clay materials used.

6.2 Containment of Wastes

On average, each household produces approximately 0.6 tonne of domestic waste each year (Department of the Environment 1991). This comprises a mixture of putrescible materials (20 %), plastics (7 %), paper/cardboard (33 %), metal (8 %), glass (10 %), textiles (4 %), dust (10 %) and other miscellaneous wastes (8 %). A similar waste stream derives from commercial premises. In addition, industry produces a vast amount of waste, ranging from inert or semi-inert soils to hazardous chemical liquid and solid wastes. In total, approximately 150 million tonnes of wastes are currently produced in the UK each year.

Although moves towards integrated waste management, comprising a combination of waste minimisation, recycling, pre-treatment (e.g. incineration, composting, anaerobic digestion) and finally disposal, have moved towards the top of political agendas in recent years (e.g. Department of the Environment, 1995b), the vast majority of wastes generated are still disposed of in landfill sites, and even with an integrated waste management strategy there will always be a similar requirement for disposal of waste residues. At present, approximately 124 million tonnes of waste are disposed of at over 4000 licensed waste disposal sites in the UK. Similarly, in many other countries throughout the world (e.g. North America, Australia, South Africa, southern Europe), landfill remains the main disposal option for the majority of wastes, and even where extensive pre-treatment of waste is undertaken (e.g. northern Europe), landfill remains important as the ultimate disposal point.

Biodegradable waste materials will decompose by biological processes. The rate of degradation will vary from the rapid decomposition of vegetable matter through to the relatively slow degradation of cellulose in paper and cardboard. Other waste materials will similarly decompose, with even the majority of plastics being the subject of biological and/or chemical activity; only a few waste types (e.g. polythene, glass, inert soil) can be considered as effectively "inert". This biodegradation of waste leads to the generation of landfill gas (Department of the Environment 1989), a combination of methane (~ 60%) and carbon dioxide (~ 40%) with trace quantities of a wide range of longer-chain hydrocarbons which contribute to a distinctive aromatic odour. This gas can represent a hazard owing to the following effects:

1. methane is explosive in air at concentrations of between 5 and 15%;
2. the presence of carbon dioxide or methane displaces air with a resultant deficiency of oxygen which may cause asphyxiation;
3. similarly, the deficiency of oxygen in soil owing to the presence of carbon dioxide or methane can cause vegetation distress or dieback.

The infiltration of water through degrading wastes causes leaching of the by-products of the decomposition process to produce a liquor which is termed "leachate". In the early stages of decomposition a strong 'acetogenic' leachate (see below) is generated (Robinson and Gronow 1993). As decomposition proceeds, landfills typically turn 'methanogenic' as different bacteria become active, with a resultant weaker leachate with the composition shown in Table 6.1.

Table 6.1. Typical acetogenic and methanogenic leachates

	Acetogenic	Methanogenic
pH	6.5	7.5
Chemical oxygen demand, mg/l	30000	2000
Biological oxygen demand, mg/l	15000	400
Ammoniacal nitrogen, mg/l	1000	1000
Chloride, mg/l	2000	2000

Both of these leachates are strong effluents with, for example, considerably higher concentrations of contaminants than a typical sewage. The chemical and biological oxygen demands can lead to large reductions in the amount of dissolved oxygen in any receiving waters with a resulting detrimental impact on aquatic flora or fauna. Ammonia is a List 2 substance, as defined by the EC Groundwater Directive (1990), and is a toxic material which can have a detrimental effect on water as a resource at only low concentrations. Chloride too can impair the quality of water supplies when present in relatively small amounts.

Clearly there is a need to protect the environment from the potentially detrimental impacts which could result from the migration of landfill gas or the leakage of leachate from wastes disposed of in landfill sites if these are not controlled (Fig. 6.1). This is generally achieved by a combination of the following (Fig. 6.2):

1. the controlled extraction of landfill gas by actively pumping from wells within the landfill followed by flaring and/or utilisation (e.g. electricity generation), thus reducing the potential for off-site migration;
2. the provision of a leachate drainage system to collect and remove leachate from the landfill before it is able to leak into the surrounding environment;
3. the provision of a lining system, isolating the wastes from the surrounding environment;
4. the provision of a capping layer to minimise the infiltration of rainwater into the landfill, hence reducing leachate generation, and preventing the uncontrolled egress of landfill gas to the atmosphere (methane is a particularly harmful greenhouse gas, the uncontrolled emission of which is no longer supported (Department of the Environment 1995a).

The latter (provision of a lining system and a capping layer) require the use of low-permeability materials. Sometimes synthetic materials (e.g.

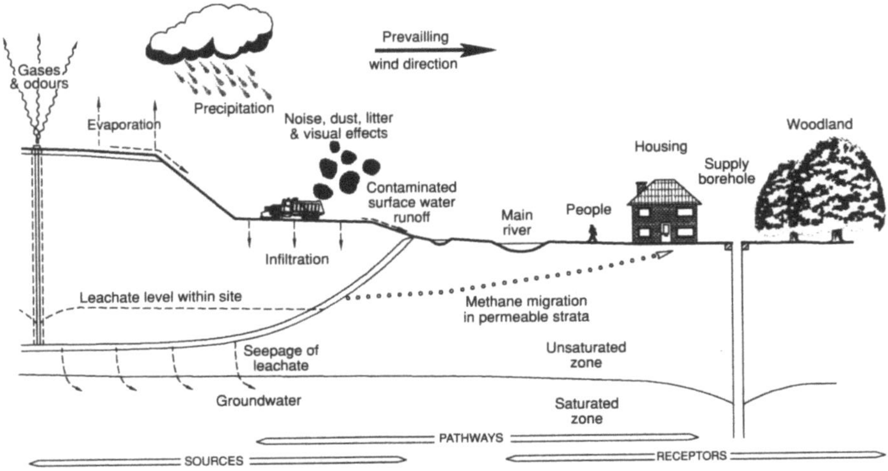

Fig. 6.1. Schematic source/pathway/receptor diagram for an area containing landfill. (After Department of the Environment 1995a)

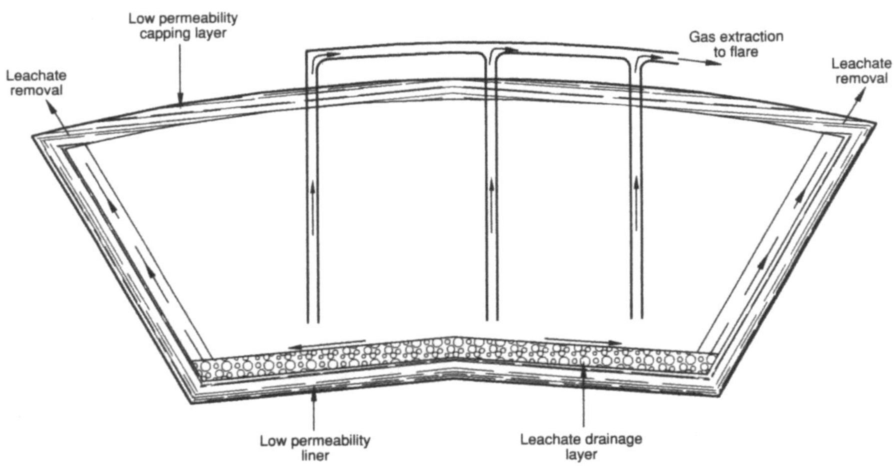

Fig. 6.2. Schematic diagram of leachate and gas management systems within a landfill

geomembranes, hydraulic asphalt) are used, but whenever they are available, naturally-occurring, low-permeability materials (i.e. clays) can form an appropriate alternative as a partial or total replacement for synthetic materials. The benefits of using clay in this context and the physical and chemical properties which make it suitable for use in landfills are described below.

6.3 Why Use Clay?

The reason for using clay in landfills is essentially a combination of design and accident. Historically, landfill sites were just that: typically old 'pits' which were infilled with waste literally on a village-by-village basis. These 'pits' were generally former mineral workings, for example chalk, gravel or brick (clay) pits or stone quarries. When a scientific approach was applied to this ad hoc practice (e.g. Department of the Environment 1978), it soon became clear that those waste disposal sites situated in high permeability ground such as gravel, fractured rock or chalk often caused some of the environmental problems described above. However, sites located in clay pits were generally secure and were observed to contain both leachate and landfill gas.

In the 1980s, there followed a deliberate policy in the UK of often situating new landfill sites in former clay pits or, where sites were located in higher permeability ground, using clay as a basal liner. Where water pollution presented a problem associated with an existing landfill site, a common solution was to place a clay capping layer over the site to minimise rainwater infiltration and hence reduce leachate generation. Therefore, where locally available, natural clays rapidly became established as appropriate materials for use as environmental barriers around landfills, either as an in situ material or as an engineered liner or capping layer. The main advantages of using clays can be summarised as follows.

1. *Cost.* Where clays are locally available, either on or close to the disposal site, they represent the most economic solution to providing landfill containment.
2. *Low Permeability.* Owing to their small particle size, clays exhibit a very low permeability (Sect. 6.4.1) with very low rates of seepage of the fluids which are present in landfills.
3. *Robustness.* Thick clay barriers are generally less prone to damage than their thinner synthetic counterparts.
4. *Availability.* The widespread distribution of clays throughout the UK (see below) ensures that low-permeability materials are present close to the majority of waste disposal sites.
5. *Chemical Compatability.* Clay minerals represent the end-product of geological weathering over a period of often thousands of years, and are thus largely inert and generally not subject to significant chemical reaction with leachate (although see discussion below).

Low-permeability clays are present in most regions of the UK. Over the Midlands, north Wales, northern England, Scotland and northern Ireland, boulder clays associated with recent glaciations are present as a veneer above bedrocks, some of which [e.g. Keuper Marl (Triassic), Carboniferous mudstones and marls] themselves weather down to a low-permeability clay soil. The younger strata which are typically prevalent in southern and eastern England [e.g. London Clay (Eocene), Lias and Oxford Clays (Jurassic), Gault Clay (Cretaceous)] are often unlithified and can be excavated with ease to produce a clay material which can be used for landfill containment. All of these materials are frequently used as barriers in landfills.

Where clays are not available in close proximity to landfill sites, their use can be uneconomic. However, sands or coarse silts can be used in combination with 'synthetic' clays, termed bentonite, to produce a barrier with similar low permeabilities to natural clays. Bentonite is a generic term for calcium (Ca) or sodium (Na) montmorillonite. This material exists as a weathered residue of igneous rocks in some parts of the Mediterranean (Greece and Italy), the United States (Wyoming) and to a lesser extent in the UK (e.g. Fuller's Earth contains Ca-montmorillonite). These clays have hygroscopic properties such that, when wetted, they swell up to three times their original volume for Ca-bentonite and up to 15 times their original volume for Na-bentonite. The raw materials are milled into powder or granular form, then mixed with sands/silts, either by rotovation or preferably by batch mixing; upon wetting, the bentonite swells up to fill the void spaces between the sand/silt particles thus forming a low-permeability mineral liner, termed bentonite-enhanced sand (BES). Bentonite may also be used as part of a synthetic liner, termed a Geosynthetic clay liner (GCL), where the bentonite in powder or granular form is sandwiched between two geotextile layers, held together by sewing, needle punching or glueing (Koerner et al. 1995).

In recent years, environmental concerns have led to an increase in the standard required from landfill liners, both in the UK and throughout the developed world, such that at least two levels of protection are provided. In the event of the failure of one of these levels, the second has to be capable of providing adequate protection to the environment. These two levels of protection can be a single-liner system with pre-defined remedial actions in the event of unacceptable rates of leachate seepage or landfill gas migration occurring. However, more often than not, at least two independent barriers forming a composite or multiple-liner system (Department of Environment 1995 a) are used. It has also been recognised that the respective properties of synthetic (geomembrane) liners and clays can be mutually beneficial when used in combination with the performance of the whole exceeding that of the sum of the parts (Thomas 1991). This has led to the prevalence of liner systems which comprise a geomembrane overlying, and in intimate contact with, a mineral liner (clay, BES or GCL). This has actually led to an increase in the use of mineral liners, such that virtually every new landfill in the UK will contain clay, in some form, as an integral part of its liner system.

6.4 Properties of Clay

6.4.1 Hydraulic Properties

The main property of clay relevant to its use as part of an environmental barrier around landfills is its low permeability. If clay is in situ, it is important that its permeability is sufficiently low to prevent seepage of leachate or migration of landfill gas where it would have a significant environmental impact; if clay is to be engineered, it is important that it can be placed and compacted to attain and retain this low permeability.

In the landfill industry the term permeability is generally used synonymously with the term hydraulic conductivity. This is not a purist approach, with intrinsic permeability having an absolute value, independent of the permeant, whereas hydraulic conductivity relates solely to the flow of liquids (usually water); in landfill sites the permeant is known to be a liquid (leachate) and the simplified terminology is generally used.

In order to ascertain the seepage rates through clay liners with different permeabilities, a simple Darcy-based calculation can be undertaken. In reality, this calculation makes a number of simplified assumptions; for example, Darcy's Law applies to saturated flow whereas clay liners and capping layers are often situated in the unsaturated zone, and conservative assumptions are generally made in calculating hydraulic gradients. Nonetheless, this provides a useful approach for comparing seepage rates through clay liners of different permeability when different heads of leachate are applied. Such an approach (e.g. Seymour 1992) yields the equation:

$$q = \frac{k(l + d)}{d},$$

where q is the seepage rate ($m^3/m^2/s$), l is the leachate head above top of the liner (m), d is the thickness of the liner (m) and k is the hydraulic conductivity (m/s).

Applying a variety of hydraulic conductivity values to a 1 m leachate head with various liner thicknesses yields the seepage rates shown in Fig. 6.3.

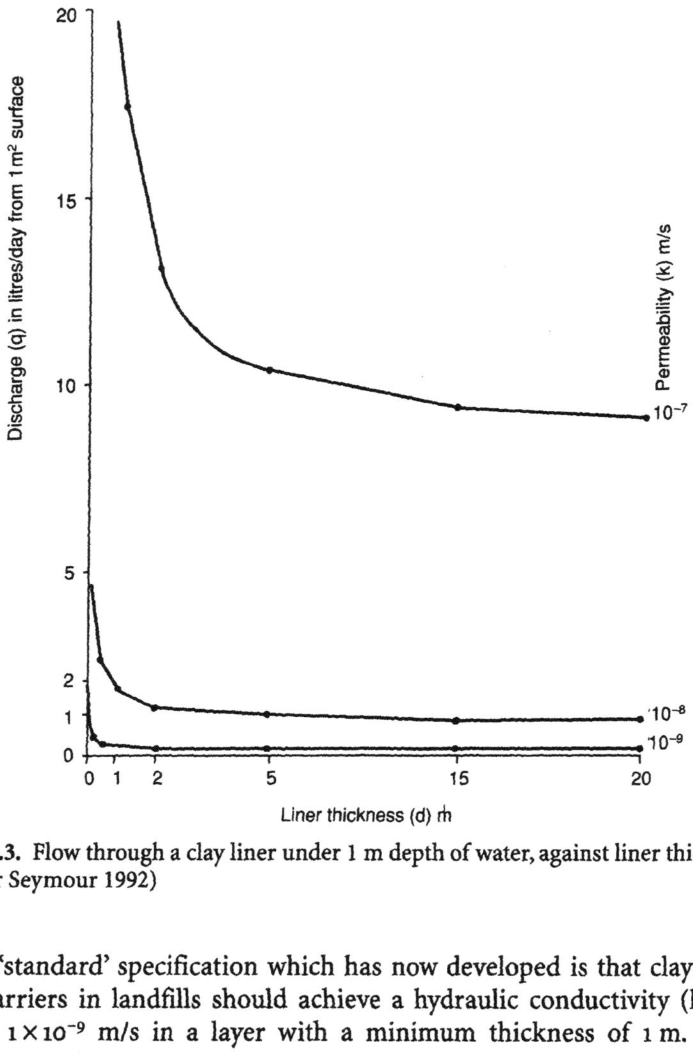

Fig. 6.3. Flow through a clay liner under 1 m depth of water, against liner thickness. (After Seymour 1992)

A 'standard' specification which has now developed is that clays used as barriers in landfills should achieve a hydraulic conductivity (k) less than 1×10^{-9} m/s in a layer with a minimum thickness of 1 m. These values are chosen for the following reasons:

1. The permeability value is generally achievable by most naturally-occurring clays provided that they are excavated, placed and compacted in a controlled engineering operation.
2. A 1-m-thick layer represents a robust barrier which is less prone to damage than thinner layers.
3. The calculated seepage rate through such a liner with a maximum 1 m leachate head (also a standard specification), even making the conservative simplifying assumptions described above, is 1.7 $m^3 ha^{-1} day^{-1}$

which represents a significant environmental benefit compared with the amount of seepage which would occur if no liner were present (equal to the amount of leachate generated = amount of rainwater infiltration = 8.2 $m^3 ha^{-1} day^{-1}$ using a typically effective rainfall value of 300 mm year). It is interesting to note that the same environmental benefit could not be demonstrated for a clay liner with $k = 1 \times 10^{-8} ms^{-1}$ (seepage rate = 17 $m^3 ha^{-1} day^{-1}$ compared with a leachate generation rate of only 8.2 $m^3 ha^{-1} day^{-1}$).

From these calculations, as summarised in Fig. 6.3, it can be seen that only small changes in permeability can have a large effect on seepage rates and hence the environmental benefit attributable to a clay landfill liner. It can also be seen that, given the large area of landfill sites (often many hectares), total seepage rates can still be relatively high, even through such low permeability materials. As such, the $1 \times 10^{-9} ms^{-1}$ specification should be considered as a minimum and, where practically achievable, efforts should be made to form a clay barrier with an even lower permeability.

6.4.2 Geotechnical Properties

In most civil engineering activities, the aim is to place and compact material to achieve a high strength and density. For barriers in landfill sites, the permeability of the material is also of paramount importance and material should be selected with this criterion in mind. The following factors may affect the permeability of a clay barrier, both during and subsequent to its construction:

1. *Particle Size Distribution.* In general terms, materials with high silt and clay contents will have a low permeability. Materials with a high percentage of stones or excessively large particles are generally unsuitable. The overall grading is also important, and it should be noted that well-graded materials with a low clay content can perform acceptably (see below).
2. *Particle Shape and Fabric.* Well-graded materials will tend to compact to a lower porosity (and hence permeability) than uniformly-graded materials. In addition, fabrics present within clay, such as bedding planes, shear zones or other discontinuities can lead to permeabilities being anisotropic (Arch and Maltman 1990).
3. *Degree of Compaction.* Well-compacted material will tend to have a lower porosity (and hence permeability). It is therefore important to ensure that the moisture content of the clay is conducive to its adequate compaction. In addition, if material is not well-compacted, then

clods may remain. Poorly-compacted clays are also more susceptible to increases in their permeability occurring after compaction.

4. *Plasticity.* In order to enable adequate remoulding to form a clay barrier, the clays used need to be sufficiently plastic. However, clays which are too plastic tend to be liable to high shrinkage/swelling as moisture content changes, both of which can affect permeability after construction.
5. *Moisture Content.* The moisture content of a clay can affect the ability to compact it which will affect the permeability. It has been found that the lowest permeability values are achieved when the clay is slightly wetter than the optimum moisture content which achieves the maximum dry density (see Fig. 6.4 after Daniel 1990). This moisture content allows compaction close to optimum conditions, but provides sufficient moisture to facilitate remoulding of materials and breaking down of clods.

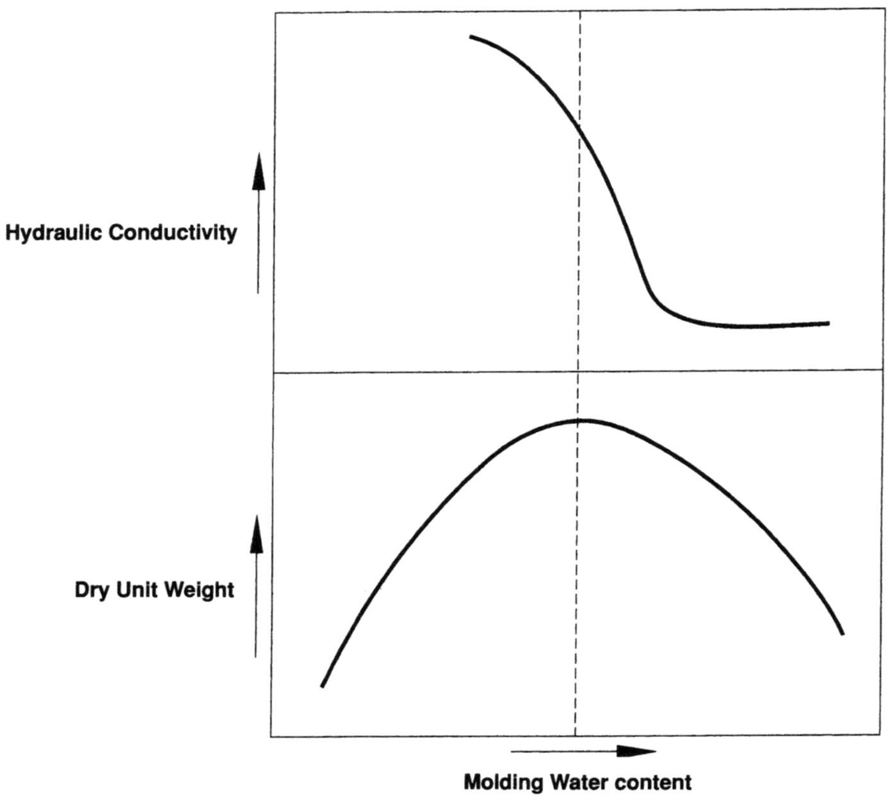

Fig. 6.4. Influence of moulding water content on the hydraulic conductivity of a compacted clay. (After Daniel 1990)

All of these factors are of importance if a clay barrier is to be constructed, and an investigation and testing programme needs to be instigated to verify that the proposed material is capable of achieving the required specification. 'Standard' specifications (e.g. *National Rivers Authority – North West Region 1989*; Daniel 1990) in terms of these parameters are given below:

Plasticity index	10–30%
Liquid limit	<90%
Clay content	>10%
Silt + clay content	>30%
Gravel content	<10%
Maximum particle size	50 mm

It should be borne in mind that this standard specification is for guidance only and is not absolute. Materials which do not meet these criteria may still be suitable and are still used as clay barriers in landfills, although more thorough investigations and trials are generally required.

6.4.3 Chemical Properties

6.4.3.1 *Effects on permeability*

Clays are chosen for use as environmental barriers around landfills because, being at the end of the geological weathering process, they are largely inert. However, early research work (summarised in Daniel and Liljestrand 1984) showed that certain substances such as acids, bases, and organic compounds could cause an increase in the permeability of clays, presumably due to chemical reaction. Other authors (e.g. Brown and Thomas 1984; Johnson et al. 1989) report that diffusive flow of organic contaminants (e.g. gasoline, kerosene) may be such as to increase the effective permeability by as much as four orders of magnitude. However, other research programmes (Finno and Schubert 1986; e.g. Ilgenfritz et al. 1988) reported no such detrimental effect with similar tests. A comprehensive study undertaken by Daniel et al. (1988) into the compatibility of clay barriers with a variety of different, more realistic landfill leachates concluded that:

> "none of the leachates produced an increase in permeability of the subsoils or a significant change in the index properties of the soils … in fact, there were two positive benefits: (1) some of the

leachates caused a gradual reduction in the permeability of the clay (in some cases to the point where the clay was practically impermeable); and (2) the leachate passing through the clay was chemically filtered by the soil such that the amount of hydrocarbon material coming out of the test specimens was far less than the amount going into the specimens".

Similarly, a study by Peters (1993) reported that:

"the permeability of natural clayey bedrocks does not seem to be affected by the leachates",

although he noted mineralogical changes in the uppermost 5 cm of one of the three clay liners he investigated due to the effects of the highly concentrated industrial waste leachate over a period of 25 years, with no mineralogical change below this depth. Similar results were reported from laboratory studies undertaken by Shimek and Herman (1985) and Wuellner et al. (1985) which showed typical landfill leachates to be chemically compatible with compacted clay specimens.

In summary, the research work undertaken to date indicates that clay liners are generally compatible with the composition of a 'conventional' leachate; however, if significant concentrations of organic contaminants are present, as may be the case at certain industrial landfills, chemical interactions may result in the performance of a clay liner being significantly inferior to that which was expected; nowadays, this is generally taken into account, with synthetic geomembranes often used as a primary liner in such sites, thereby preventing direct global contact between the leachate and the clay. A notable exception to this conclusion is bentonite, which is susceptible to cation exchange causing the calcium or sodium cations present within the bentonite to be substituted by other cations that are present within the leachate. This is of particular concern where high swelling sodium bentonite is used, where cation exchange can cause "reversion" to a more stable but less-swelling form with a corresponding increase in permeability. This is best illustrated by the difference between a bentonite hydrated with distilled water and with landfill leachate (Hookes et al. 1987) where the bentonite hydrated with landfill leachate showed a reduced swelling and a higher permeability. To minimise these effects, some manufacturers add a polymer to their bentonites to act as a retardant to cation exchange. Various studies (e.g. Grantham and Robinson 1988) have shown landfill liners formed by sand mixed with polymer-treated bentonite to be largely effective, and such bentonites continue to be used as "synthetic" clay barriers around landfills.

6.4.3.2
Effects Due to Attenuation

Many clays have actually been shown to have a beneficial impact on leachates which seep through them due to a process known as attenuation (Department of the Environment 1978). In this process, biochemical interactions cause substantial reductions in the concentration of some organic and inorganic components of leachate, and processes such as filtration, immiscibility, sorption and ion-exchange have been shown to prevent the migration of various cations, hydrocarbons and ammonia through soils and rocks, although chloride has been found to be largely unaffected.

This attenuation effect has been found to be greater with some materials than with others; in particular, bentonite has a high cation exchange capacity and has been occasionally used to attenuate contaminants. However, such practice of allowing leachate to percolate through the unsaturated zone into receiving groundwater is now rarely permitted, and it is the containment and not the attenuating properties of clays which are of upmost importance. It is worth noting, however, that the industry is now moving into adopting a risk-assessment approach to landfill development (Department of the Environment 1995a). Attenuation processes are becoming recognised as forming an important part of this approach.

6.5
Investigation and Testing

In the assessment of a particular site to determine its suitability for development as a landfill, a site investigation is invariably required. The aims of this site investigation, according to the UK Department of the Environment (1995a), are as follows:

1. to assess the ground conditions (geology and hydrogeology) under the site in order to determine the environmental impacts which would accrue were leachate to leak from the site;
2. to assess the potential for gas migration from the landfill through the substrata;
3. to determine background groundwater conditions (quality, level, flow direction and flow rate);
4. to determine background soil gas conditions (composition, concentration);
5. to determine the engineering properties, in relation to foundation stability and slope stability, of the materials on the site;

6. to assess the availability and suitability of materials on the site for use in landfill earthworks.

Investigation and testing of clays for use as landfill barriers will generally form an integral part of such a site investigation. It is therefore important to obtain representative samples of the clay material which is proposed for use. Where the clay is to be recompacted to form an engineered clay barrier, the samples may be disturbed bulk samples obtained from boreholes, trial pits or existing stockpiles. Where insitu clay is to be relied upon to provide landfill containment, undisturbed core samples are required. In addition, it is often beneficial to carry out tests in boreholes or trial pits to determine the insitu hydraulic and geotechnical properties of the clay material. Permeability tests in boreholes (falling head, packer or slug tests) can provide useful data, and insitu shear strength measurements can be done. Other in situ tests (e.g. geophysical methods) are possible, but rarely undertaken on landfill sites.

The number of samples obtained essentially depends on the variability and geometry of the source; for example a uniform 'clay pit' site would require fewer samples than a more complex site which contained a number of discrete lenses of potentially heterogeneous clay. In all cases, however, sufficient samples should be obtained to be able statistically to verify the suitability and homogeneity of the clay.

In order to investigate the suitability of the samples obtained for use as a clay barrier, against the criteria discussed in the previous section, the following tests are necessary:

1. *Determination of Natural Moisture Content.* This routine test provides information about the ease of compaction and the homogeneity of the material.
2. *Determination of Atterberg Limits.* These tests assess the plastic limits, liquid limits and plasticity indices of the material from which its plasticity characteristics can be determined.
3. *Particle Size Distribution.* In order to determine the clay, silt + clay percentages as well as the gravel/cobbles content of the clay.
4. *Determination of Compaction Characteristics.* This assesses the relationship between moisture content and dry density for a given compactive effort, and enables determination of the maximum dry density and optimum moisture content. In addition, a moisture condition value (MCV) may be used to provide a rapid determination of material suitability during construction (Murray et al. 1992).
5. *Remoulded Shear Strength.* This test is useful in assessing whether or not the clay material can practically be placed and compacted with conventional civil engineering equipment. (As a general rule,

the undrained shear strength for remoulded material should exceed 40 kN/m^2).

6. *Effective Stress.* Where clay slopes are to be left exposed for a considerable period of time, there may be slope-stability concerns which require effective stress testing to define long-term drained shear strength parameters.
7. *Permeability Tests.* These should be undertaken on both undisturbed and remoulded samples (see below).

Test methods are generally conventional, and are described in detail in BS 1377 (British Standards Institution 1990). Permeability measurements, for example, may be made in the laboratory by one of three methods:

1. Permeameter – constant head or falling head,
2. Triaxial apparatus (termed the flexible-wall method),
3. Oedometer.

The permeameter can be prone to leakage around the edges of specimens, particularly if the edges are not perfectly smooth (e.g. if stones are present). However, this will tend to overestimate permeability which, for clay barriers in landfill sites, errs on the side of caution. The oedometer test indirectly assesses permeability (k) by determination of the coefficient of volume compressibility (M_V) and the coefficient of consolidation (C_V), using the equation:

$k = C_V \gamma_W M_V$, where γ_W is the density of water.

This method is notoriously unreliable and can underestimate permeability, sometimes by as much as two orders of magnitude. This is due to the very thin specimen which is used in the oedometer test which can lead to significant errors with the sample not being representative, particularly if stones are present. In reality, the formula given above recognises this error in the oedometer, and was actually derived for determining C_V using k values derived from field tests, not the other way round. Unfortunately, the inexpensive nature of this test means that it is still often used, but the test is flawed and sole reliance should not be placed on test results.

The most accurate method available for the determination of permeability in the laboratory is the flexible-wall method in the triaxial apparatus. This involves placing a core sample in the triaxial apparatus, applying a confining pressure to seal the edge of the specimen, then determining the permeability, normally using the constant-head method. The test is done in three stages: consolidation, where the clay is allowed to consolidate in response to the applied confining pressure; saturation, where air which could cause errors in the test is driven out of the specimen and; permeability determination, where the flow rate through the specimen is measured for a given head.

In cases where insitu material is to be relied upon to act as a clay barrier, it is the in-situ permeability which is relevant and which should be determined using undisturbed samples in conjunction with field measurements. In cases where the material is to be excavated, placed and compacted to produce a new clay barrier, it is the remoulded permeability which is important. As discussed above, this can vary depending on the material's moisture content and degree of compaction, and it is important to quantify this relationship.

A suite of tests should therefore be undertaken whereby the material is compacted to varying dry densities at different moisture contents, with the permeability measured at each stage. These data are then used to define an 'envelope of acceptability' in terms of the moisture content and dry density within which the material achieves the required permeability specification (usually 1×10^{-9} m/s). If the material is highly heterogeneous, then it may not be possible to define a representative envelope of acceptability. In this case, it may be prudent to consider the material being used as a number of discrete 'sources' with a discrete envelope of acceptability defined for each source. In reality, the required extent of a testing programme to produce meaningful results may be huge; this problem is compounded by the considerable time and expense involved in accurately determining the permeability in the triaxial apparatus (cost is often several hundred pounds per specimen). Therefore, a pragmatic approach is often adopted using less data, but taking into account the results of previous studies (e.g. Daniel 1990); for example, it is generally recognised that clay should be compacted such that it achieves a maximum of 5% air voids and this should form the lower boundary of any envelope of acceptability. It is also well established that permeability increases at moisture contents dry of optimum: the optimum moisture content therefore provides a sensible lower boundary to any envelope of acceptability.

Simple field measurement can then be used during construction (see below) to determine whether or not the compacted clay falls within the envelope of acceptability. This is demonstrated by the data in Fig. 6.5 which show the actual permeability data and envelope of acceptability for a landfill site in northwest England.

Even with such simplifying assumptions, it is always important to demonstrate that the clay which it is proposed to use can be placed and compacted to meet the required permeability specification. This is best achieved by the construction of small test cells which would normally

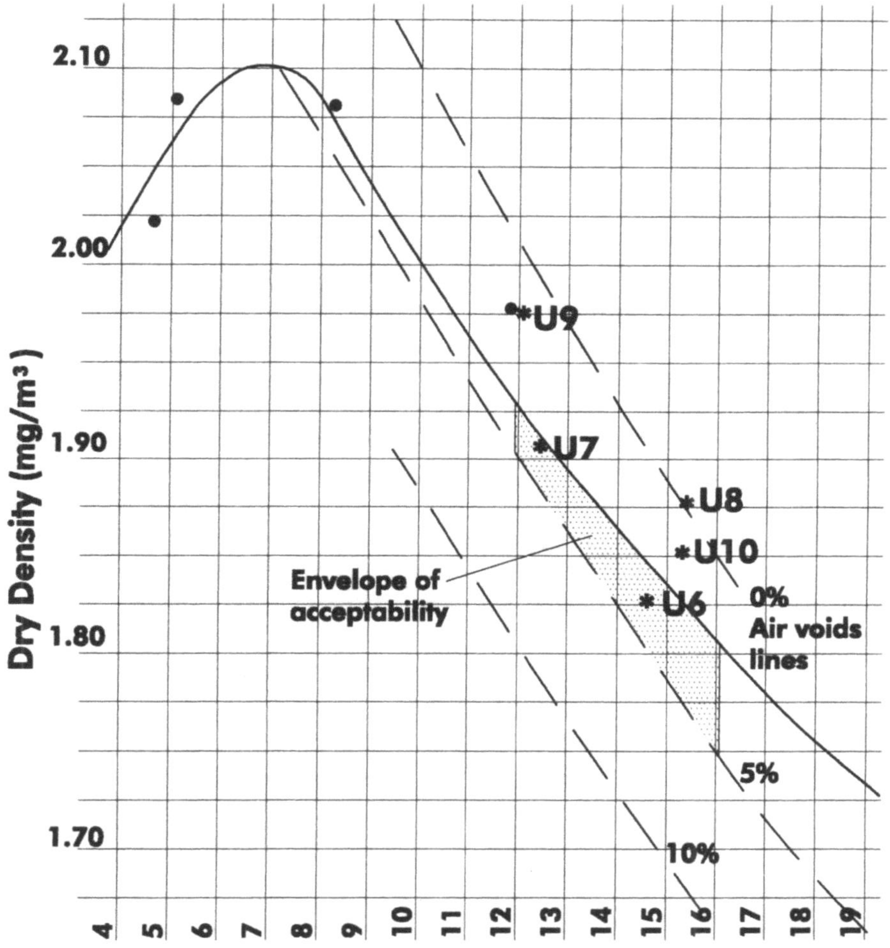

Fig. 6.5. Typical envelope of acceptability for placed and compacted clay in terms of moisture content and dry density at which appropriate permeability requirements are met

involve the construction of a trial clay barrier over a small area using the materials, equipment and personnel which it is intended to use in the construction of the full-scale clay barrier. Such test cells have the following advantages:

1. they permit verification that the material can be placed and compacted to meet the required specification;

2. they allow refinement of the method to be used in placing and compacting the clay (e.g. thickness of layer, number of passes, compactive effort) in order to optimise permeability;
3. they allow any teething problems with construction to be ironed out before full-scale construction of the clay barrier starts.

These test cells need to be thoroughly tested with the material's moisture content, dry density and the permeability of core samples all measured, to verify that the relationship between these parameters and the envelope of acceptability determined in the laboratory applies equally in the field.

Once the programme of investigation has been completed, it should be well established that the material proposed for use meets the required specification, under what compaction conditions (moisture content, dry density) the specification is met, and which construction methods (plant, layer thickness, number of passes) optimise compaction. Only if all these aspects are known should construction of the clay barrier commence. Otherwise, there is a significant risk that compaction will not be optimised, and that the compacted clay will not meet the required permeability specification, needing reworking (removal/replacement) with associated costs.

6.6 Construction: A Case Study Illustrating Quality Control and Quality Assurance

6.6.1 Introduction

Historically, clay barriers in landfills have been poorly constructed with very little control. As such, some of the first clay barriers constructed around landfills did not meet their required objectives and have resulted in environmental problems. Nowadays these problems are avoided by a programme of construction quality control (CQC) and independent construction quality assurance (CQA) aimed at ensuring that:

1. the material being used is relatively homogeneous and is similar to that which was characterised as part of the laboratory testing programme;
2. the methodology (i.e. layer thickness, plant used, number of passes) follows that determined by construction of the test cells;

3. the moisture content and dry density fall within the envelope of acceptability;
4. the permeability meets the required specification.

These criteria are met by defining the programme of construction, inspection and testing in three documents:

1. Construction method statement: which defines the plant, personnel, workmanship and materials to be used.
2. CQC plan: which defines the programme of inspection and testing to be undertaken in order to check that the required specifications are met.
3. CQA plan: which defines the programme of independent observations and tests to verify that construction has followed the method statement, that the CQC plan has been followed and that test results demonstrate compliance and, if necessary, independent tests to check those from the CQC programme.

It is now a normal requirement for such procedures to be followed in the construction of clay barriers in landfills, and many regulatory authorities, insurers and financiers require independent certification that the barrier has been constructed, inspected and tested properly, and that the required specifications are met. Various authors (e.g. Daniel and Koerner 1993) have provided data about the anticipated performance of clay barriers with or without such independent CQA. These accounts show considerable benefits to be achieved through a comprehensive approach to supervision of clay barrier construction with resultant 'average' permeabilities lower by as much as two or even three orders of magnitude.

The exact contents of method statements, CQC and CQA plans will vary on a site specific basis, depending on the nature and source of the material, the geometry of the clay barrier, the plant available etc. Similarly, the parties responsible for drafting and implementing the respective documents will vary depending on the specific contractual arrangements and regulatory requirements for any particular site. However, as an illustration, a summary of the contents of these documents for a successfully completed project undertaken by Aspinwall & Company involving the construction of a clay liner on a quarry slope is listed below with plates included to illustrate how construction was actually achieved.

6.6.2
Construction Method Statement

6.6.2.1
Introduction

This section describes the construction procedures to be followed by the earthworks contractor in the construction of the liner system. It should be read in association with the construction details shown on the drawings.

Inspection and testing and independent quality assurance procedures are described in the CQC plan and the CQA plan.

The following details represent the intended method of working at present, based on the results of previous construction experience. However, it should be appreciated that site conditions and operational considerations may require minor changes to these procedures being instigated at short notice, although the general principles will remain unaltered. Any such changes will only be implemented by agreement between the earthworks contractor, the landfill operator and the CQA Consultant. Appropriate addenda will be appended to this method statement as soon as is practicable following the agreement of any such changes.

6.6.2.2
Plant and Personnel

The following plant will be used in the construction of the liner system:

1. 360° Excavator(s),
2. Dump trucks,
3. Loading shovel/dozer,
4. D6-type dozer,
5. 6.5-tonne smooth-wheeled/vibratory roller.

The following personnel will be used in the construction of the liner system and capping layer:

1. Machine operatives,
2. Earthworks foreman,
3. Labourers,
4. CQC technician.

The construction works will be supervised by the supervisor, and overseen by the landfill manager.

6.6.2.3
Site Formation

The formation level has been formed by quarrying operations. The pre-existing quarry vertical slopes have been cut back to an angle of 20° (1 in 2.75) to facilitate lining. Site formation levels will then be verified by profiles/travellers with an allowable tolerance of ±25 mm.

Prior to lining, a 100-mm blinding layer of silt/clay will be placed on the prepared cut slope and proof rolled by a 6.5-tonne smooth-wheeled roller, 2 m in width, towed by a dozer to form a firm base for the liner system. Any soft or wet areas will be either left to dry out before being rolled, or will be excavated and replaced by acceptable material placed in accordance with the requirement of the Department of Transport's *Specification for Highway Works: Series 600 – Earthworks*, 1991.

Inspection and testing procedures associated with the site formation, prior to the placing of the clay, are described below.

6.6.2.4
Material Selection

Clay material shall be excavated from the stockpile using a 360° excavator and loaded into dump trucks. This material will be tested, as described below, with adjustments to the moisture content made as appropriate.

These vehicles shall then transport the clay material directly to the area where the liner system is being constructed. There will be no stockpiling of clay material prior to its placement. As far as is practicable, these vehicles shall not track directly onto the prepared site formation or on compacted layers of clay, although a certain amount of such tracking is inevitable.

Any damage so caused to the prepared site formation or to the compacted clay shall be made good by rolling and tracking prior to further layers of clay being placed.

6.6.2.5
Placing of Clay Layers

The transported clay will be bladed out using a D6-type dozer to form a layer 310 mm thick (see Fig. 6.6a).

The track pattern of the dozer shall be grousers to ensure breaking down of clods. In the event that large clods remain following placement, these will be broken down using the wheels of the landfill compactor.

a

b

Fig. 6.6a. Placement of a clay liner on a quarry slope

Fig. 6.6b. Compaction of clay liner by driving forwards and backwards along each strip of placed clay the required number of times

c

d

Fig. 6.6 c. Compaction testing of clay barrier

Fig. 6.6 d. Completed clay barrier

Profiles will be constructed adjacent to each strip as it is constructed and used in association with travellers to verify the thickness of each layer.

6.6.2.6
Compaction of Clay Layers

Each 310-mm layer of clay will be subjected to a total of six passes by a 6.5-tonne vibratory roller, 2 m in width, towed by a dozer. The compaction will be achieved by driving forwards and backwards along each strip of placed clay the required number of times (see Fig. 6.6b).

Following compaction of each layer, tests will be carried out as described below.

Prior to the placing of further clay layers, the surface of the compacted clay will be roughened using dozer tracks in order to provide an effective key with the overlying layer.

The final layer of clay will be proof rolled with a smooth-wheeled roller to form a flat top to the lining system. The tolerance of this level shall be +25 mm/−10 mm.

The 1-m liner will therefore be formed from four separate layers, each placed at a thickness of 310 mm, compacted down to 250 mm. The total thickness will be verified by profiles/travellers once the liner is complete, as described below.

6.6.2.7
Sequence of Construction

The clay liner will be placed in two strips each approximately 60 m long and 20 m wide (i.e. extending all the way up the slope). These "strips" are shown on the drawings.

The construction of the clay liner will be phased with strip A constructed before strip B. A stagger of at least 2 m will be left between each layer to allow for adequate compaction at the edge of each strip.

Placing and compaction of clay will not take place in conditions of heavy rainfall or if the clay is wet and unworkable.

6.6.2.8
Protective Layer Placement

Once the clay liner has been placed, it will be immediately covered by a 300-mm protective layer of inert construction/quarry waste with a maximum particle size of 100 mm.

6.6.3
Construction Quality Control Plan

This section describes the CQC procedures which will be followed by the Earthworks Contractor in their inspection and testing of the liner system.

This section should be read in conjunction with the CQA plan which defines the CQA procedures to be followed by the CQA consultant.

Records will be kept of all inspections and tests undertaken in association with this work, as described below.

6.6.3.1
Definitions

CQC is defined as those actions which provide a means to measure and regulate the characteristics of an item or service to contractual and regulatory requirements. In relation to this particular project, this refers to inspections and tests undertaken by the earthworks contractor to ensure that the materials and construction methods are conductive to a finished installation which meets the requirements of the site licence.

6.6.3.2
Personnel

CQC testing will be undertaken by the Technical Manager of the earthworks contractor. Laboratory testing will be undertaken at the CQA laboratory.

6.6.3.3
Site Formation

Once the site formation, including the blinding layer, has been prepared and rolled, a visual inspection will be carried out to check for tolerances and soft spots.

Any problem will be identified and corrected – uneven areas by regrading, and soft spots by recompaction or removal/replacement. These areas will be re-inspected prior to placement of the clay.

6.6.3.4
Materials

Records shall be kept of the source of all materials used in the construction of the liner system.

A routine moisture content test shall be undertaken on the clay stockpile being used at the start of each working day, or following periods of weather which could significantly change the moisture content of the material during the working day. This shall be determined by weighing material prior to and after its desiccation in a microwave oven.

6.6.3.5
Placing of Clay Layers

Profiles will be constructed around each strip of clay. These will be used in association with a traveller to verify that the placed thickness of clay is 310 mm. Any unevenness will be smoothed out using a dozer and the thickness rechecked prior to compaction of each layer.

Once each layer has been placed a visual inspection will be carried out to ensure that clods of clay have been sufficiently broken down. If large clods remain, these will be broken down further, either by tracking by a dozer or, if necessary, tracking by the landfill compactor. This inspection will also identify whether any erroneous materials are present within the placed clay. Affected areas will be re-inspected prior to compaction of each layer.

6.6.3.6
Compaction of Clay Layers

Records will be kept of the number of passes to which each layer is subjected.

Following compaction of each layer, tests shall be carried out along each compacted strip at a spacing of approximately 30 m (equivalent to two tests per strip). These tests shall comprise dry density measurements which will use the sand replacement method in accordance with the procedures defined in *BS 1377: Part 9: 1990 – Method 2.1.* Voids formed in undertaking these tests will be backfilled by clay material and recompacted. Material excavated in the undertaking of the tests will be tested for moisture content using the microwave moisture content test (see Fig. 6.6c).

The results of these tests will be used to determine whether the compacted clay falls within the envelope of acceptability.

If the material falls outside this envelope, then further tests will be undertaken to define the extent of the problem area. This area will then be remedied as appropriate, either by further compaction or adjustment of moisture content followed by further compaction or by removal and replacement of the affected area. All such problem areas will then be retested to ensure compliance.

The thickness of each compacted layer will be checked using a traveller in association with the profiles constructed around each strip. Additionally, the total thickness of the clay liner/capping layer will be similarly verified, and a visual inspection will be made of the completed proof-rolled liner/capping system to check for unevenness, erroneous materials and desiccation cracks.

Following completion of the liner, two undisturbed core samples will be obtained (one on each strip). The permeability of these samples will be determined in accordance with *BS 1377: Part 6, Test 6* (British Standards Institution 1990), using a constant head of 1 m and an effective stress of 100 kPa, in order to verify that the required hydraulic conductivity values have been achieved. Voids will be backfilled with recompacted clay or bentonite pellets.

6.6.3.7
Documentation

Records will be kept of all CQC inspections and tests using the CQC forms.

Record plans shall be kept showing the locations of placed strips and tests.

6.6.4
Construction Quality Assurance Plan

This plan addresses the Construction Quality Assurance (CQA) monitoring of the construction of a lining system.

The CQA monitoring will be undertaken by the CQA consultant to supplement and augment the CQC checking undertaken by the Earthworks Contractor. As such, this CQA plan should be read in association with the construction method statement and CQC plan described above. In addition, this CQA plan includes, consolidates and integrates the requirements of the waste management licence.

6.6.4.1
Definitions

CQA is defined as a planned and systematic pattern of all means and actions designed to provide adequate confidence that items or services meet contractual and regulatory requirements and that the works are undertaken in accordance with the required specifications. In relation to this particular project, this refers to procedures employed to assure conformity of the materials and construction methods to this CQA Plan, the construction method statement, the working plan and the conditions of the site licence to assure optimum performance of the liner system.

6.6.4.2
Parties and Responsibilities

The Operator of the landfill is ....

The Earthworks Contractor involved in the construction of the liner system and capping layer is .... The Contractor is responsible for the construction of the works with associated CQC.

The CQA Consultant is .... The CQA Consultant is responsible for implementing the requirements of this CQA Plan, for supervising the works and for certifying that the liner system and capping layer meet the requirements of the site licence. The CQA consultant shall appoint a CQA manager who shall be suitably experienced and qualified. The CQA manager shall be ....

The CQA Laboratory is a laboratory, independent of the Operator, the Contractor and the CQA Consultant. The CQA Laboratory is responsible for undertaking CQA testing, as instructed by the CQA Consultant.

The Waste Regulation Authority (WRA) is responsible for enforcing the conditions of the site licence.

6.6.4.3
Duties of the CQA Manager

The CQA Manager is directly responsible for the CQA of the liner system. His or her duties are:

1. to keep records of all on-site CQA activities;
2. to ensure the ongoing preparation of record drawings;
3. to assign locations for CQC inspection and testing;
4. to verify the calibration and condition of testing equipment;
5. to prepare a weekly report including relevant information generated by the CQC procedures;
6. to review the results of all inspections and testing, making appropriate recommendations;
7. to compile a photographic record of construction;
8. to identify and remedy any unresolved deviations from the CQA Plan;
9. to prepare the final CQA report.

The role of the CQA Manager is, therefore, to audit the construction work and CQC inspections and testing undertaken by the Earthworks Contractor, thus providing an extra level of checking for compliance with the requirements of the site licence. The site licence also requires the CQA Manager to have a supervisory role in the construction of the liner system and capping layer. The CQA Manager will therefore be responsible for liaising with the Operator, the Earthworks Contractor and the WRA to assure the smooth completion of the project in accordance with the requirements of the site licence.

6.6.4.4
Materials

The CQA Consultant shall verify that the material used in the construction of the liner system is derived from one of the approved locations shown on the drawings, and the CQA Consultant shall verify that the protective layer meets the required specification.

6.6.4.5
Site Formation

The Contractor will be responsible for preparing the site formation, including the blinding layer, undertaking inspections for unevenness and soft spots as part of his or her CQC Plan.

The audits undertaken by the CQA Consultant will include checks to ensure that:

1. lines and grades have been surveyed;
2. the surface to be lined has been rolled and compacted so as to be free from irregulations, protrusions, loose soil or abrupt changes in grade;
3. there is no area excessively softened by high water content;
4. the inspections required by the CQC Plan have been undertaken, and necessary remedial actions have been completed and re-inspected.

6.6.4.6
Clay Placement and Compaction

The Contractor will be responsible for placing and compacting the clay material in layers, as defined in the construction method statement. The Contractor will also be responsible for inspection and testing of the placed and compacted clay, as defined in the CQC Plan.

The CQA Consultant will verify that:

1. the clay is being placed in layers 310 mm thick forming strips with a width of approximately 20 m across the entire width of the area to be lined or capped;

2. clods of clay have been sufficiently broken down prior to compaction;
3. each layer of clay is subjected to six passes with a 6.5-tonne vibratory roller, 2 m in width;
4. the compacted thickness of clay is 250 mm;
5. the density and moisture content of the placed and compacted clay cause it to fall within the envelope of acceptability;
6. the placed material is free from boulders and other erroneous materials;
7. the compacted liner is free from desiccation cracks;
8. all inspections and tests required by the CQC Plan have been undertaken, any problem areas have been isolated, and necessary remedial actions have been completed and retested.

In addition to the auditing of the CQC Plan, the CQA Consultant will obtain samples from the placed and compacted liner and send them to the CQA Laboratory for independent testing.

6.6.4.7
Protective Layer Placement

The Contractor will be responsible for placing the protective layer, as defined in the construction method statement and for inspection as defined in the CQC Plan.

The CQA Consultant will undertake audit checks to verify that:

1. no plant tracks on the placed clay layer;
2. the thickness of the protective layer is 300 mm;
3. the inspections required by the CQC Plan have been undertaken.

6.6.4.8
Documentation

An effective CQA programme depends on recognition of all construction activities that should be monitored, and on assigning responsibilities for the monitoring of each activity. This is most effectively accomplished by the documentation of quality assurance activities. The CQA Manager will document that all quality assurance requirements have been addressed and satisfied.

Standard reporting procedures will include the preparation of a weekly report by the CQA Manager which, at a minimum, will include:

1. Contractor's CQC records;
2. field notes;
3. completed CQA proformas, as presented in the previous CQA document;

4. a running list of incomplete items and actions taken to remedy incomplete items.

6.6.4.9
Final Report

Following completion of lining and capping works, a final report will be prepared by the CQA Consultant which shall include:

1. an outline of the project,
2. a description of the constructed liner system/capping layer,
3. reference to the CQA Plan and other documents used,
4. a summary of activities,
5. a photographic record of construction,
6. material specifications,
7. copies of all CQC records,
8. copies of all weekly logs,
9. record drawings,
10. discussion of problems and solutions,
11. certification statement.

These documents define the ideal approach to be followed during construction of a clay barrier. In reality, as construction proceeds, certain aspects of the construction methods will inevitably need to change in response to circumstances, and an interactive and iterative approach is therefore needed. Such changes were required on the project described above due to unforeseen circumstances (e.g. machine breakdowns, inclement weather, unexpected heterogeneity of clay material). Nonetheless, the comprehensive nature of the CQA programme allowed the changes to be successfully implemented (see Fig. 6.6d), with the actual measured permeability of the clay liner being very close to that anticipated from the pre-construction investigation programme.

6.6.5
Aftercare

Once a clay barrier has been constructed and shown to meet the required specification, it is often assumed that the clay has attained the appropriate standard and will retain that same standard for the design life of the site. As such, the rigorous supervision procedures employed during construction are generally discontinued. However, it is important that a barrier's low permeability is maintained in order that its environmental performance is not impaired; a number of routes may result in changes to the permeability of an installed clay barrier as follows:

1. accidental damage,
2. vandalism,
3. neglect (e.g. desiccation, wetting or even vegetation growth),
4. chemical compatibility with leachate (discussed above),
5. deformation.

Accidental damage can clearly dramatically affect the integrity of any seal with a resulting impact on the environment. Indeed, its is the author's experience that damage caused to liner systems following their construction can far outweigh any benefits attributable to comprehensive CQC/CQA programmes during construction. The following recent 'accidents' have all been known to result in damage to liner systems:

1. careless tipping of bulky difficult wastes (e.g. reinforcing bars, lighting columns or concrete blocks) directly onto the liner;
2. tracking of vehicles over the liner;
3. driving of posts supporting litter netting through a liner;
4. debris from blasting operations in an adjacent quarry falling onto a completed liner.

The potential number of similar 'accidents' which must have occurred is enormous and emphasises the need for a rigorous approach to be employed up to and including the placement of the first layer of waste onto a clay liner and the placement of a soil layer above a clay capping layer.

Similarly, vandalism can cause immense damage to liner systems which can only be dealt with by enhanced security combined with the rigorous ongoing approach discussed above.

Neglect may occur when too large an area of liner is prepared and not covered with waste for long periods. This can result in desiccation or erosion of the clay barrier or simple weathering (Miller et al. 1992). Similarly, if capping layers are neglected and not maintained following their construction, damage may accrue. If clay liners are left for a period of time, vegetation may also become established, with roots having a detrimental effect on the clay.

Finally, deformation to which a clay barrier may be subjected following its construction may alter its permeability. Arch et al. (1996) describe 'dynamic' environments in waste disposal sites at the base of the site due to foundation settlement, the side-slopes of site due to slope instability, or at the top of the site due to settlement of the wastes. They report experi-

a

b

Fig. 6.7 a. Vegetation dieback on clay capping layer due to landfill gas migration

Fig. 6.7 b. Erosion gullies on completed capping layer due to neglect mental data to demonstrate that this deformation can result in increases in the permeability of a clay barrier by as much as two orders of magnitude.

A recent case study (Arch 1995) described below emphasises the importance of aftercare to maintain a clay barrier following its construction. A clay capping layer was constructed over poorly compacted waste, and was subject to comprehensive CQC/CQA procedures which demonstrated that it achieved a permeability of approximately 10^{-11} m/s immediately following its compaction. Soil was placed above the clay, and seeded to produce a grass sward which quickly became well-established. Over the next 18 months, the poorly compacted waste settled by up to 7 m. Over a similar timescale, certain areas of the grass sward died (see Fig. 6.7a), and monitoring showed that this was associated with migration of landfill gas through the capping layer. The spatial distribution of the vegetation dieback showed it to be restricted to those areas where the maximum settlement had occurred or those areas where there was a change in slope gradient where the stress was concentrated. In one area, the settlement was so pronounced that it caused the capping layer to completely shear. This suggests that the permeability in these areas had increased due to deformation associated with the settlement. In addition, heavy rainfall led to large erosion gullies developing in some parts of the capping layer which were unsightly (see Fig. 6.7b).

These aftercare issues which may impair the performance of a clay barrier following its construction require an ongoing commitment to regular routine inspections, monitoring and appropriate maintenance should defects be detected. This emphasises the need for adequate financial provisions to deal with anticipated problems as well as any unexpected events.

6.7 Conclusion

This review has described the environmental benefits which can be accrued by the use of clay barriers in landfills. It has emphasised that this represents an effective way of containing wastes in a landfill and has stressed the need for thorough investigation of the material to be used prior to its use, comprehensive CQC and CQA procedures to be followed during and after construction and maintenance of the clay barrier during an aftercare period.

Even when such a rigorous approach is adopted, pitfalls may occur which need to be dealt with. Many of these can be predicted and taken into account in design and construction, but the unexpected will always occur which emphasises the need for ongoing maintenance and a need for adequate financial provision.

References

Arch J (1995) Remediation of slope instability at an industrial landfill. Proc 5th Int Landfill Symp, 2–6 Oct 1995, Cagliari, Sardinia

Arch J, Maltman AJ (1990) Anisotropic permeability and tortuosity in deformed wet sediments. J Geophys Res 95:9035–9045

Arch J, Stephenson E, Maltman A (1996) Factors affecting the containment properties of natural clays. In: Bentley SP (ed) Engineering geology of waste disposal. Engineering Geology Spec Publ No 11, the Geological Society, London

British Standards Institution (1990) British standard methods of test for soils for civil engineering purposes. BS 1377, Parts 1–9, London

Brown KW, Thomas JC (1984) Conductivity of three commercially available clays to petroleum products and organic solvents. Hazard Waste 4:545–553

Daniel DE (1990) Keys to successful design, construction, testing and certification of soil liners for landfills. GRCDA J Munic Waste Manage 1:17–27

Daniel DE, Koerner RM (1993) Quality assurance and quality control for waste containment facilities, USEPA Technical Guidance Document, National Technical Information Service. US Dep Commerce, Springfield, Virginia, 305 pp

Daniel DE, Liljestrand HM (1984) Effects of landfill leachates on natural liner systems. A report to the Chemical Manufacturers Association, January 1984

Daniel DE, Liljestrand HM, Broderick GP, Bowders JJ (1988) Interaction of earthern liner materials with industrial waste leachate. Hazard Waste, Hazard Mater 5:93–108

Department of the Environment (1978) Cooperative programme of research on the behaviour of hazardous wastes in landfill sites. HMSO, London

Department of the Environment (1989) Waste management paper No 27. Landfill gas. HMSO, London

Department of the Environment (1991) Waste management paper No 28. Recycling, HMSO, London

Department of the Environment (1995a) Waste management paper No 26B. Landfill design, construction and operational practice. HMSO, London

Department of the Environment (1995b) Making waste work: a strategy for suitable waste management in England and Wales, HMSO, London

Department of Transport (1991) Specification for highway works. HMSO, London

EC directive on protection of groundwater against pollution caused by certain dangerous substances (1990). EC Directive 80/68/EEC, Commission of the European Communities, Brussels

Finno RJ, Schubert WR (1986) Clay liner compatibility in waste disposal practice. J Environ Engin 112:1070–1084

Grantham G, Robinson HD (1988) Instrumentation and monitoring of a bentonite landfill liner. Wastes Manage 78:134–148

Hookes J, Glas H, Koftamp J, Ryliner A (1987) Bentonite liners for the isolation of waste disposal sites. Waste Manage Res 5:93–205

Ilgenfritz EM, Blanchard FA, Masselink RL, Panigrahi BK (1988) Mobility and effects in liner clay of fluorobenzene tracer and leachate. Groundwater 26:22–30

Johnson RL, Cherry JA, Pankow JF (1989) Diffusive containment transport in natural clay: A field example and implications for clay-liner waste disposal sites. Environ Sci Technol 23(3):340–349

Koerner RM, Garung E, Zanzinger H (eds) (1995) Geosynthetic clay liners. Proc Int Symp, Nurnberg, Germany, 14–15 April 1994. Balkema, Rotterdam

Miller CJ, Lee JY, Menna J (1992) Effects of freeze-thaw on clay cover liner performance. In: Usmen, Acar (eds) Proc Mediterranean conf on Environmental geotechnology, Cesme, Turkey, 25–27 May 1992. Balkema, Rotterdam
Murray EJ, Rix DW, Humphrey RD (1992) Clay linings to landfill sites. Q J Engin Geol 25:371–376
National Rivers Authority – North West Region (1989) Earthworks on landfill sites. NRA (NW), Warrington
Peters T (1993) Chemical and physical changes in the subsoil of three waste landfills. Waste Manage Res 11:17–25
Robinson HD, Gronow J (1993) A review of landfill leachate composition in the UK. Proc 4th Int Landfill Symp, 11–15 Oct 1993, Sardinia
Seymour K (1992) Landfill lining for leachate containment. J Inst Water Environ Manage 6:389–396
Shimek SJ, Hermann DJ (1985) Effect of acidic leachate on clay permeability of clay soils. Proc 8th Annu Madison Waste Conf, Madison, 18–19 Sept 1985, pp 303–314
Thomas RL (1991) Quality control in containment engineering. J Inst Waste Manage 81(11):665
Wuellner WW, Wierman DA, Koch HA (1985) Effect of landfill leachate on the permeability of clay soils. Proc 8th Annu Madison Waste Conf, Madison, 18–19 Sept 1985, pp 287–302

Chapter 7

Clay Minerals and Health

J.C. Wagner, K. McConnochie, A.R. Gibbs, F.D. Pooley

7.1 Introduction

The health of humans can only be compromised by clay minerals which have been pulverised and dried. The respiratory organs are affected only when the amount of dust produced is excessive. Teleologically, the development of the human lungs has produced mechanisms which cope with the relatively small amounts of mineral dusts which are found in rural and most urban atmospheres. It is only when these mechanisms become overwhelmed by the quantity and toxicity of the dust that damage will occur.

The vast majority of clay dusts are innocuous, and therefore only produce deleterious effects if the exposure is intense. A few react through contaminants, but we have only found one group, the palygorskites, in which there is evidence that the minerals themselves are capable of causing damage to the respiratory system.

We have investigated the biological effects of exposure to the following clay groups:

1. Kaolinites,
2. Palygorskites,
3. Vermiculites,
4. Mica,
5. Bentonite (montmorillonite).

We feel that these materials cover a wide spectrum of the clays to which people may be exposed. The reasons for studying these clays in the first instance were requests either from Government agencies or Industry to assess the possible hazards which could occur if humans were exposed, and to establish acceptable atmospheric concentrations if required. We have found the most satisfactory method of approach is by team work, with teams of scientists with varying scientific expertise and equipment investigating the problem on a constant collaborative basis. With sufficient financial backing we approach the problem in the following manner.

Assessment by:
1. Mineralogists, industrialist, and mining engineers.
2. Clinical medical teams with expertise in industrial medicine, including radiology, pulmonary physiology, and epidemiology.
3. Pulmonary pathologists with specialised knowledge.
4. Experimental pathologists with specialist training in the respiratory systems of laboratory animals, working with dust physicists who have developed techniques for the exposure of animals to specific types and consistent concentrations of the dusts to which humans have been exposed.

More sensitive cellular and molecular methods of study are available, e.g., tissue culture techniques, which are required for the more detailed study of reaction of dust and tissue, particularly in the study of mechanisms.

This chapter describes the manner in which the health effects of specific mineral particles have been investigated utilising a combined disciplinary approach, with emphasis being placed upon exposure to clay minerals. The biological effects of heavy dust exposure in animal experiments are discussed with human exposure experience to similar mineral dusts and the type of disease patterns this has produced. The importance of mineralogical techniques and information in establishing the aetiology of dust-related diseases is outlined, illustrating the multidisciplinary nature of research in this area.

7.2 Experimental Pathology

Experimental studies, designed in relation to dusts and the lungs, establish whether or not the dusts can evoke fibrosis or neoplastic change in the exposed animals. Two main types of studies are undertaken: inhalation studies in which mineral dusts are inhaled directly, and in-vitro studies in which dust particles are brought directly into contact with cell cultures. Obviously the most physiologically relevant method of exposure is by inhalation but this method is extremely expensive, and can only be undertaken in a few centres. Another method is direct injection into the respiratory system, but this may induce an acute toxic reaction. To test for the carcinogenecity of fibrous dusts and to answer the question "If the dust does reach the pleural or peritoneal cavity can it cause malignancy?"

two types of in-vitro studies are used – intrapleural and intraperitoneal implantation.

Rarely are such experiments carried out for purely scientific purposes. There are often political implications in understanding the biological effects of the dusts, with trade unions, industry, and legal constraints playing a part.

7.2.1 Materials

The dust to be tested should taken from a sample of material that is actually going to be in commercial use, and it is pointless to use a dust that will never occur in industry. From both human and past animal experience we can now narrow the range and the size of the particles or fibres. Particles which are possible to inhale should be in the region of 5.0 µm diameter spheres, while fibres should be less than 3.0 µm in diameter, and indeed a diameter of less than 0.25 µm is accepted as significant for the production of mesotheliomas. The essential length is between 5.0 and 30.0 µm.

Both positive and negative controls should be used. Oregon erionite is a useful positive control [because it produces mesothelial tumours so readily in rats, special precautions must be taken in handling it (Wagner et al. 1985)], while a glass wool of more than 1.0 µm in diameter and 10–20 µm in length is probably the best negative control.

The results of the experiments apply only to the dust actually used and should not be applied to other material however closely related.

7.2.2 Experimental and Test Systems

The problem with the obvious and much used animal, the rat, is that it is subject to pulmonary infections which have a two-fold effect. They cloud the issue as to how much of the pathology is due to the dust, and they shorten the rat's life; longevity is needed to study neoplastic changes, should they occur. The main pulmonary infection is caught during weaning, so caesarean-derived rats should be used which are 'specific-pathogen-free', after which they must be 'barrier-maintained', that is isolated from other rats. At death, a post-mortem of their lungs must be examined by histological methods to ascertain that their lungs are free of infection. Rats are suitable however, as they are capable of contracting pulmonary fibrosis, carcinoma of the lung and mesotheliomas. Their disadvantage is that the experiments will take at least three years (Wagner et al. 1973).

Another animal that has been used, although less frequently, is the hamster. Again infection may play a part in these animals so that a colony must be known to be free from infection. Less is known about these animals, and in particular the effect of inhalation of dusts. Mice, likewise, have been used for intrapleural, intraperitoneal, and intratracheal studies, but not for inhalation. Sheep have also been exposed to dusts by bronchoscopic canulation, with a slow infusion into one of the lobes of the lung (Begin et al. 1983).

Whatever animal is used, a full post-mortem must be carried out, including histology of all relevant organs. The histological changes observed in the sections of the lungs of rats, following exposure to dusts, are graded on a scale of 1–8, grade 1 being normal, 2–3 the reaction to the dust, and 4–8 progressive severity of the fibrosis. Grade 8 is thus a completely fibrosed lung (Wagner et al. 1974).

Many different in-vitro test systems are used, and the majority measure either cytotoxic effects, or alterations in the cell systems that might contribute to neoplastic change.

7.2.3 Experiments on Specific Clay Minerals and Groups

7.2.3.1 *Kaolinite*

In earlier years, between 1948 and 1978, tracheal instillation and peritoneal injections were used (Wastiaux and Daniel 1990).

Kaolinite was found to be mildly fibrogenic.

Inhalation studies. There are two reported inhalation studies, both using Cornish Clay, and both using rats. In the experiment by Wagner (1990), 32 Fischer 344 Specific Pathogen Free rats, four to each cage, were exposed when the rats were 47–67 days old. The respirable concentration in the dust clouds was 10 mg/m^3, and the dust clouds were generated for 7 h/day for 5 days every week for a period of 12 months. Two grades of China Clay were used, a filler grade, 92% respirable kaolinite, and a coating grade, which was 95%. Rats of each sex were used and were killed at 3, 6, 12, 24, and 28 months. Following a full necropsy the sections were examined blind of both the exposed animals and the unexposed controls. Even at 24 months with either of the kaolinite dusts, only minimum tissue reaction was noted (mean reaction). There were neither carcinomas of the bronchus nor mesotheliomas induced.

In the experiments reported by Wastiaux and Daniel (1990), 48 female rats were exposed to 300 mg/m³, that is 30 times the dose of the previous experiment. Exposure was for 6 hours per day, 5 days per week for three months. Eight animals were killed at 3, 6, 12, 24 and 28 months. At three months there was an acute reaction to the dust, which by 12 months had developed into a severe reaction with some scarring. The difference in these two experiments obviously lies in the dosage, and it has to be said that such a large dose is unrelated to human exposure levels.

In Vitro studies. Cytotoxicity studies (Vallyathan et al. 1988) were compared with the reaction following intratracheal installation. Rat alveolar macrophages were incubated with different mineral dusts for 2 h at concentrations which ranged from 0.1 to 2.5 mg/ml. Haemolytic activity was measured using sheep red cells (Harington et al. 1971). Kaolinite was more haemolytic than silica, a known haemolytic agent, at all concentrations of the dust, and selective release of lysozomal enzymes in the macrophage system was also higher. However, in the in-vivo experiment, 3 months after exposure, the rats exposed to kaolinite had virtually normal lungs, whereas those rats exposed to silica had interstitial granulomas, which persisted even at 6 months.

7.2.3.2
Palygorskites

The group in these studies included mineral samples which were labelled attapulgite, sepiolite, and palygorskite itself.

Attapulgite. Intrapleural and intraperitoneal: Stanton et al. (1981) reported the introduction of two samples of palygorskite from Attapulgus, Georgia, into the pleural cavity, at a dose of 40 mg (shorter than 5 µm, and consistently small in diameter) and produced 2/29 sarcoma/mesotheliomas. In 1981 Pott injected palygorskite intraperitoneally. This was reported together with results on sepiolite and palygorskite in 1989 (Pott et al. 1990). Four samples of palygorskite from four different locations were tested: Caieres (Spain), Georgia (USA), Lebrija (Spain) and Mormorian (Finland). Ten mg of dust was given intraperitoneally. Dust 1 contained 240 fibres $\times 10^6$ per mg (length 5 µm) and produced 12/30 sarcomas/mesotheliomas. Dusts 2, 3, and 4 had long fibres as follows 0.61, 0,34 and 0.06 $\times 10^6$-fibres over 5 µm respectively. For these last three dusts, 60 mg was given intraperitoneally. They produced four tumours out of 112, 113 and 114 rats (3.6, 3.5, 3.5% respectively). Accepting that up to 10% of mesotheliomas may occur in control rats, it would therefore suggest that only the dust from Caieres would be dangerous, and agrees with the number of long fibres present. Wagner et al. (1987) using the intrapleural method, used a sample from Lebrija and one from Torrejon el Rubio (Spain). Ten mg were injected. All fibres were found to be less than 2 µm in length for the Lebrija sample and 2/38 mesotheliomas were produced (one peritoneal). The specimen from Torrejon contained 0.85×10^6 fibres/µg. They were more than 6 µm in length, and less than 0.5 µm in diameter. It produced 14/26 mesotheliomas. This palygorskite was exclusively used for drilling-mud consolidation when drilling oil wells in the North Sea.

Inhalation
Only the Lebrija sample was tested (there was insufficient material from Torrijon). At post-mortem there was no evidence of fibrosis (grade 3). One peritoneal mesothelioma and one adenocarcinoma were produced.

Sepiolite. Intrapleural and Intraperitoneal: Wagner et al. (1987) used two samples of Madrid sepiolite, a crude sample directly from a drill core which was used both ultrasonically dispersed and undispersed. The other sample was from a commercial product. There were 15.77×10^6 fibres/µg, and no fibres of either greater than 0.5 or 0.2 µm in diameter or longer than 6 µm were present in each sample respectively. The ultrasonicated crude sample and the commercial product produced 1/40 mesotheliomas, while the other crude sample produced none. Pott et al. (1990) used a sepiolite from the same source in Madrid, and also a second source from Finland. For the first sample he inoculated 80 mg intraperitoneally, introducing 180×10^6 fibres/mg (of length greater than 5 µm). There were 2/32 sarcomas/mesotheliomas. With the Finnish fibre only 10 mg was used in the same type of inoculation, but there were 5500×10^6 fibres (same length measurements) per mg. This resulted in 24/36 similar tumours. This Finnish sample is not used commercially.

Inhalation
Wagner et al. (1987), using the inhalation method as described for Kaolinite, produced only a grade 3 (early interstitial reaction), no mesotheliomas, two squamous carcinomas, two bronchial carcinomas, two bronchiolar alveolar hyperplasia (one associated with one of the carcinomas), in a group of 40 rats. These findings were taken as within the expected range for these rats.

Palygorskite. Intrapleural: Wagner et al. (1987) reported on a sample of palygorskite that had been obtained from a disused quarry in Leicester in the English Midlands. Ten mg was again injected intrapleurally per rat.

There were 191.9×10^5 fibres/µg of respirable dust, with 36.5×10^5 greater than 6 µm in length and 34.7×10^5 fibres 0.5 µm in diameter. This dust produced 30/32 mesotheliomas.

Inhalation
The regime was the same as with the sepiolite. There was a grade 3 reaction (early interstitial, but no fibrosis). There were two pleural and one peritoneal mesotheliomas, one malignant alveolar tumour, and eight bronchiolar alveolar hyperplasia. This material is not now exploited for commercial use.

In the I.A.R.C. (International Agency for Research on Cancer) (1987) monograph on the evaluation of carcinogenic risk of chemicals to humans (Vol. 42), the information on palygorskite and sepiolite is reviewed. The final evaluation for palygorskite is that: "There is limited evidence for the carcinogenicity of palygorskite to experimental animals. There is inadequate evidence for the carcinogenicity to humans". For sepiolite the final evaluation was "There is inadequate evidence for the carcinogenicity of sepiolite in experimental animals. No data were available to evaluate the carcinogenicity of sepiolite in humans".

7.2.3.3 *Vermiculites*

We could only find reports of cytotoxic studies.

7.2.3.4 *Mica*

There is one report by Sahu (1990) from India of exposure of mice, rats and guinea pigs to the muscovite variety of Indian Mica, from the Bihar province. Samples of the mica dust were made (Zaidi 1969) with particles sized below 5 µm.

Intraperitoneal. Mice were inoculated as follows: Twenty mice were inoculated with 5.0 mg/0.1 ml saline, and 20 further mice were used as controls, 0.00 mg/0.1 ml. The mice were examined at 24 h, 3, 7, 15, 30, 60 and 90 days.

Intratracheal. Mice, rats and guinea pigs were given single doses. Eighty mice were given 5.0 mg/1.0 saline, with 40 controls, while 10 rats were given 25 mg/1.0 saline, and 36 guinea pigs with 4 controls were given 75 mg/1.5 ml saline. Mice were examined at 7, 15, 30, 60, 90, 120, 150 and

210 days. Rats were examined at 90, 180 and 330 days. Guinea pigs were examined at 24 h, 7, 15, 30, 60, 90, 120, 180, 300 and 365 days.

The results of the peritoneal injection in the mice showed, after 90 days, that macrophages packed with the dust were present in lymph nodes, and there was early tissue reaction in the peritoneum.

After the intratracheal inoculation in mice, there was an immediate acute inflammatory reaction, while by 210 days post-inoculation, the lung lesions had shrunk. Much of the dust had left the lungs and was in the lymphoid tissue, with some resultant fibrotic reaction there. Rather similar results were found with the rats and guinea pigs.

7.2.3.5
Bentonite (Montmorillonite)

Vallyathan et al. (1988) studied bentonite in the same experiment as that for kaolinite already mentioned. The response in vivo was again an acute inflammatory response, and like kaolinite but unlike silica the response had completely resolved in 6 months after exposure. Like kaolinite, bentonite was found to be more haemolytic in vitro than silica, but bentonite, unlike kaolinite, inhibited the activity of alveolar enzyme activity.

In Vitro Tests. A review of these short-term tests has been carried out by Jaurand and Pairon (1990). They point out that comparison between one mineral and another was difficult because the number of particles for a given weight of mineral was not comparable.

Most of the experiments which are comparable have been carried out on red blood cells using haemolysis as an end point. Kaolinite, bentonite, montmorillonite, palygorskite, and sepiolite were toxic to red blood cells, thus causing haemolysis. Micas have little effect on the same system.

Vermiculite was tested by Vallyathan et al. (1988), and found to be as haemolytic as bentonite, kaolinite and silica. The degree was less when standardised for differences in surface area.

7.3
The Clinical Effects of Clay Mineral Inhalation

7.3.1
Clinical Methods

Clinical medicine has two roles in the field of occupational lung disease. First there is the need to use clinical techniques for epidemiological studies, and secondly, the occupational physician is able to offer clinical advice to incapacitated individuals. This latter role, whilst important, will not be further considered here. Instead, the methods used by clinicians to gather epidemiological information will be described, followed by a brief look at the results of some epidemiological studies.

The major tool used by the clinician to investigate a potential pneumoconiosis (occupational lung disease) are as follows:

1. the collection of information including details of past and present health, smoking habits and occupational exposure to dusts and fumes;
2. a physical examination, which is of lesser importance in epidemiological studies than in clinical care of an individual;
3. an X-ray examination of the chest;
4. measurement of the functioning capacity of the lungs; and
5. the collection of mortality information.

Epidemiological studies are broadly of two kinds; the more common cross-sectional study which looks at a workforce at one point in time, and the less common but more useful longitudinal study which looks at a group of workers at one point in time and then at the same group either some years later or follows them until death. Longitudinal studies are particularly useful when they follow up not only the workers in question, but also a similar group of people (a control group) who have not been exposed to the hazard under investigation. The two groups can then be compared to assess their levels of ill-health and their causes of death to see if there are real differences between them which could have been the result of the workplace exposure.

The clinical history. The collection of information about an individual is called the clinical history. Questionnaires are often used to collect this information as their use helps to iron out discrepancies. Even so, it has to be appreciated that it is not possible to do this completely. For example, if the questionnaire is administered to 100 hypothetical workers, it is quite possible that 20 of them will affirm that they cough regularly. If the same questionnaire is repeated one week later, it would not be a surprise if 25 report a regular cough and 1 week later still perhaps 19 will report a regular cough. It is not that those answering questionnaires set out to confuse, it is just a fact that a group of people do not always give the same answer to the same question when asked at intervals. Analyses of information from questionnaires must therefore take this into account.

Another problem area is cigarette smoking. Smokers who are trying to give up will often report erroneously that they have been successful or they will not admit to their smoking habit for fear that it will be used as an explanation of their symptoms. Conversely, it is unusual for nonsmokers to say that they smoke so that positive smoking histories are usually an underestimate.

A final problem concerns memory. It may be quite difficult for a 60-year-old man or woman to remember all of his/her working history, particularly if he/she moved frequently from job to job.

It is for these reasons that the use of a control group can be so useful. The control group must be carefully chosen so that it is similar in all respects to the study population, apart from occupational exposure. It is likely that the same discrepancies will occur in each population, making it easier to see real differences which may be attributable to the hazard concerned. For example, if 75 + 5% of the workers report a regular cough compared to 25 + 5% of the controls, then it would be worth considering that something in the working environment causes a cough. Using sophisticated statistical techniques this argument can be extrapolated and expanded.

The physical Examination. This is rarely of use in epidemiological studies of respiratory disease, unlike the study of occupational skin disease where it may be very useful.

The X-ray Examination. This comprises an ordinary chest radiograph though in some situations it is not enough and more specialised images have to be obtained. A radiograph is a picture. It can be described in words but is much more useful if it can be described numerically. This is possible using a technique known as the International Labour Organisation Classification of Chest Radiographs (International Labour Office 1980). A detailed description of the method can be found in any clinical textbook concerned with occupational lung diseases. In essence, it converts the radiographic appearance into a series of numbers and letters (a, b, c). This means that the appearance of each radiograph can be reduced to numbers and letters which relate to the amount of abnormality present. It is therefore possible to construct an aggregate score for each population studied and these scores can be compared, thus any differences between, say, a dust-exposed working population and their control group would immediately become apparent.

It is also possible to use these radiographic scores to construct dose-response curves. Within a population exposed to dust there will be those with a heavy exposure and those with less. Estimates of dust inhaled over time can be set against the amount of radiographic abnormality. Within the limits of the error of all biological measurements, it is possible to make reasonably accurate predictions of how much dust will cause how much radiographic abnormality (which with many industrial exposures relates to abnormal lung pathology – see before). For example, it can be estimated that a certain level of respirable dust in the atmosphere will result in 1% of those exposed developing a significant radiographic abnormality.

The construction of these dose–response curves has been used in many industries to set levels of acceptable atmospheric dust – though what is acceptable is more often an economic and political question than a medical one. As an extreme example, politicians in a drought-ridden area which requires the services of a potentially hazardous industry to bring in water which will save many lives may accept that 10% of the workers in that industry will develop pneumoconiosis, whereas the same industry in a different area may aim for a pneumoconiosis rate of less than 1%.

Economics also affects decisions. It is often very expensive to reduce dust levels, and such costs are reflected in the price and hence the market competitiveness of the industry concerned.

Lung Function Measurements. These are often used in both cross-sectional and longitudinal surveys. Tests range from those that are extremely simple and inexpensive to those that are more complicated, requiring equipment costing several thousands of pounds. In general terms the simpler tests are more reliable. Their interpretation is much more accurate in the hands of those familiar with the pitfalls. In a few situations, the lung function tests are the main indicator of abnormality, the chest radiograph being normal. More often, changes in lung function tests are related to changes on the chest radiograph.

Mortality Information. Longitudinal studies are more powerful than cross-sectional studies. The best studies of all are those that follow a working population after retirement until death and compare their life expectancy and cause of death with those of their controls. This is particularly important where the hazard in question has a delayed effect on the lung.

This is of necessity a brief account, and readers are directed to the texts of Parkes (1994) and Hunter (1994) for more detail.

7.3.2 Epidemiological Studies

7.3.2.1 *Kaolinite*

Inhalation of sufficient quantities of respirable size particles of kaolinite has been shown to cause abnormality on the chest radiographs of exposed workers, and to cause an associated abnormality of lung function (Sheers 1964; Lapenas et al. 1984). Kennedy et al. (1983), in a cross-sectional study for the USA, showed impairment of lung function associated with severe dust retention. Most recently, Rundle et al. (1993) reported a cross-sectional UK study where all present and retired china clay workers were invited to take part in a health survey. Of current employees, 70% took part compared with 17% of retired workers. The survey consisted of chest radiographs, lung function tests, and a questionnaire on respiratory symptoms and smoking habits. Only 0.8% of this population (lower than previous studies) showed significant radiographic abnormality. Conditions, i.e. dust concentrations, have improved in the industry from the 1970s onwards, and it was thought that this accounted for the general improvement in china clay workers' health found in the survey.

7.3.2.2 *Palygorskites*

Attapulgite. In a cross-sectional study reported in 1988, Gamble et al. (1988) looked at a cohort of 717 palygorskite workers from the USA. They used questionnaires, lung function tests and chest radiographs. They found no consistent association of symptoms and amount of dust exposure. There were low levels of radiographic change but there was no assocication between these and changes found in the lung function tests. The overall conclusion was that there was no consistent association of respiratory morbidity with exposure. However, the authors suggested that surveillance should continue to generate longitudinal data.

Sepiolite. There have been no longitudinal studies of sepiolite workers. In a cross-sectional study of Madrid sepiolite exposure, McConnochie et al. (1993) found mild reductions in lung function tests with length of exposure (20 years +) at a level often found after exposure to dust with no known biological effects, often referred to as nuisance dust. There was no association with the radiographic appearances.

7.3.2.3 *Vermiculite*

Amandus et al. (1987) have undertaken a comprehensive survey of USA vermiculite workers. They found increased mortality from non-malignant respiratory disease, though this showed no relationship to exposure.

They also discovered an increase in lung cancer which did show a relationship to exposure. However, the vermiculite ore to which these workers had been exposed was contaminated by amphibole asbestos, and so it was impossible to tell whether the changes seen were due to vermiculite or due to asbestos. McDonald et al. (1988) looked at a different USA population where the vermiculite had minimal asbestos contamination and found no adverse effects. Likewise, Hessel and Sluis-Cremer (1989) reported a cross-sectional study of South African vermiculite workers where there was little asbestos contamination. They found no excess of respiratory symptoms; lung function was comparable to a control group as were radiographic changes. Whilst these authors found no evidence of disease, they suggested that both longitudinal studies and mortality studies should be conducted.

7.3.2.4
Mica

There have been case reports of mica contributing to both chronic bronchitis and simple pneumoconiosis (Davies and Cotton 1983). As with vermiculite, the contamination of mica has been thought to be the cause of any changes found. A recent case report has cast some doubt on this. Landas and Schwartz (1991) described a 63-year old man with a long history of exposure to mica who had evidence of lung damage and abundant mica in his lung tissue. Asbestos and other silicates were not found. It may be that for some reason this man had a particularly heavy exposure. There have been no cross-sectional or longitudinal studies of cohorts of workers.

7.3.2.5
Bentonite (Montmorillonite)

Early reports suggested that fuller's earth may produce a relatively benign form of pneumoconiosis rather like that produced by kaolinite (Sakula 1961). It is, however, recognised that bentonite deposits often contain silica, a well-known cause of lung damage, and so it is not clear in some cohorts which mineral is to blame. A case reported by Gibbs and Pooley (1994) on a subject with pneumoconiosis who had been exposed to a Wiltshire deposit of montmorillonite, clearly showed that the lung changes were caused by montmorillonite in the absence of silica. More studies of this type are required to clarify the roles of the different minerals in the different geographic areas of fuller's earth production.

7.3.3 Summary

The information available about the clinical effects of clay on the lung suffers from several drawbacks. Firstly, there is a lack of longitudinal and mortality studies. There are many reasons for this, not least of which is the dedication required of one research group to continue to follow a population over many years, tracking down those who slip through their net. Funding can sometimes be a problem as can access to the workers concerned. Not every management group values research.

Secondly, many clays are contaminated by other materials which are known to be toxic to the lung, thus making the role of clay less certain. Elucidating the separate roles of clay and its contaminants requires the skills of the experimental pathologist and mineralogist in both human and experimental studies.

Nonetheless, it does appear that there is no serious burden of respiratory ill-health associated with exposure to these substances, and such as there is can be substantially reduced by attention to industrial hygiene. Hopefully, in the future, there will be more reports of longitudinal studies to substantiate this opinion.

7.4 Pathological Effects of Exposure to Clays

In general, significant pathological changes to the lung only result from prolonged heavy exposures to clays (in excess of 20 years), since they are weakly fibrogenic. We are concerned only with inflammatory and fibrotic responses since exposures to clay are not associated with the development of malignancies unless they are contaminated with other minerals such as amphibole forms of asbestos (McDonald et al. 1986; Amandus and Wheeler 1987). Contamination of the clays by other minerals such as quartz may also be a problem in assessing the effects of particular clay exposures. A useful way to sort out this particular problem is to look at the lung tissues of individuals who have been exposed to the particular mineral in question and who have come for post-mortem and then to make a careful assessment of the pathology and characterise the dust content of the lung mineralogically. There have been a number of studies which have utilised this approach and which have clarified the roles of the particular clay of interest and the associated non-clay minerals in producing the clinical, radiological and pathological changes (Wagner et al. 1986; Gibbs 1990; Gibbs and Pooley 1994). There is a good deal of evidence to indicate that in most circumstances it is the minerals which are retained within the lungs which are responsible for any disease.

7.4.1 Pathology of Lung Tissue Exposed to Clay Dusts

The majority of the changes seen in the lung following exposure to clays show a commonality and are not specific for each clay mineral. These have been described for kaolinite (Wagner et al. 1986), mica (Gibbs 1990) and montmorillonite (Gibbs and Pooley 1994). In addition, there are other changes which tend to be associated with certain specific minerals.

The initial changes are seen using light microscopy and are not visible macroscopically. These consist of accumulation of dust particles both free and within macrophages in the vicinity of the walls of the respiratory bronchioles, (this is the site for the deposition and retention of all mineral dusts). This is accompanied by little or no fibrosis. As the lesion enlarges, some fine strands of fibres become evident and this mixture of dust laden macrophages, dust particles and reticulin can be seen to extend from around the respiratory bronchioles to involve the adjacent alveolar ducts and alveoli. These lesions correspond to grey–white macules up to a few millimetres in size, visible macroscopically. This will correspond with small irregular opacities on chest radiographs. In the majority of cases the lesions do not progress beyond this and there is little functional consequence. In a small proportion of cases, where exposure has been very long, and there has been a very high retention of dust within the lung tissues, usually greater than 2% by mass, lesions may link up with others to produce a more diffuse pattern of fibrosis. This fibrous tissue is mixed with large quantities of dust particles and dust laden macrophages. This may be associated with severe architectural distortion of the lung and with a honeycomb pattern macroscopically. If this is extensive within the lung then this will result in respiratory symptoms, abnormalities of lung function and more extensive changes visible on the chest X-ray. This diffuse pattern of interstitial fibrosis can result in the death of the individual. On occasion, for reasons we do not understand, large mass lesions (progressive massive fibrosis) develop which can measure several centimetres in diameter. These are physically soft, grey–white, relatively homogeneous lesions which may cavitate. Circumscribed nodules are not a feature of clay pneumoconiosis, and if seen should alert one to significant concomitant exposure to quartz.

Other less consistent features that are seen include ferruginous bodies (Gibbs 1990), and foreign body giant cells. These have been more commonly observed in those exposed to mica than to the other clays. No descriptions of the pathology in humans are available for palygorskites and vermiculite.

7.5 Mineralogy of Clay Dust Exposures

The biological reactivity of specific mineral dusts and their capability of precipitating lung-related disease has historically been anecdotal. Employing only the working history of individual diseases cases has often resulted in the implication of only the major mineralogical component of a past dust exposure. The total mineralogical composition of the dust to which past exposure has occurred may not always be the most important factor in the aetiology of the disease. Of most significance are those mineral particles in the dust which are capable of penetrating to the gas exchange regions of the lung, where they may be retained.

Retention can be related to excessive exposure to the dust, the morphology of individual particles and enhanced biological reactivity of specific minerals. These factors can result in an accumulation of particles which may induce a range of adverse biological responses.

To accurately determine the cause of a dust related disease and link it to a specific mineral requires that a detailed mineralogical analysis of the dust content of diseased lung tissue is performed.

The importance of the identification of material of non-biological origin that may be found in human lung tissue specimens, obtained from diseased and control cases, has in recent years received increasing attention. This has arisen because of the demonstrable association of diseases such as cancer, with the inhalation and retention of dust particles formed from certain mineral species such as zeolites and fibrous amphiboles (Wagner et al. 1960; Baris et al. 1987).

7.5.1 Analytical Methods

The historical and detailed investigation of the dust content of the lung tissues is a long one which is closely linked to the development of analytical techniques. Wet chemical techniques and optical microscopy were the methods initially employed. However, in recent years development of sophisticated analytical techniques and instrumentation has enabled data of a physical and chemical nature to be collected from single often submicron sized dust particles (Pooley 1977).

Optical microscopic techniques are used extensively by pathologists to study the anatomy of diseased and normal tissues and to locate deposits of dust in the lung. These techniques are not, however, useful for identifying or quantifying the composition of very small samples of dust material. Some estimation of morphology and number of particles in tissue can be made using these procedures, but only of those particles that are large enough to allow their optical properties to be determined.

Wet chemical techniques are elementally specific and of little use as a means of identifying and quantifying individual specific mineral phases in lung tissue. They find some use when estimates of heavy metals are required and the investigation of specific elements in tissues such as beryllium are undertaken.

X-ray diffraction techniques have been found to be very useful for the identification of the mineral constituents of dust samples extracted from tissue samples (Nagleschmidt 1956; Gibbs et al. 1992). They have been employed extensively to monitor the presence of specific minerals such as quartz in dust extracts from the lungs of workers with a history of employment in the minerals industry, such as workers involved in the production of kaolinite and china stone in Cornwall (Wagner et al. 1986).

The techniques that have been employed in such investigations include the use of both X-ray powder cameras and X-ray diffractometry, the latter being a more useful procedure but with a limited application due to the problem of the size of samples available for examination.

Electron optical techniques of analysis which include the use of both the scanning (SEM) and transmission electron microscopes (TEM) have found an extensive use in the analysis of dust samples. The TEM has been found to be the most useful equipment to be employed in the analysis of dust samples, especially when it is combined with suitable X-ray analytical equipment to provide chemical information concerning the particles examined. The chemical information from single particles can serve as the basis for an identity of individual minerals contained in samples, while the analysis of large numbers of particles individually and simultaneously provides the basis for the determination on a quantitative basis for the mineralogical composition of dust samples. The application and use of the analytical techniques outlined and the preparation procedures necessary to prepare dust extracts from tissue specimens have been extensively investigated (Pooley 1981).

7.5.2 Dust Characteristics of Clay Mineral Exposures

Exposure to dust consisting of mixtures of clay minerals is common worldwide, and occurs mainly in agricultural areas. It is only in occupational situations that exposure to dust which is predominantly mono-mineralic occurs. Environmental dust consist of a mixture of mica related minerals, kaolinite and variable amounts of quartz, rutile and iron oxide particles. Cases of pneumoconiosis as a result of such environmental exposures have been recorded (Fennerty et al. 1983; Norboo et al. 1991) although it has not been possible to ascribe the aetiology of the disease observed to clay mineral particles alone.

Occupational exposure to relatively pure clay mineral dusts are rare and associated mainly with the mining of deposits of such minerals or where copious amounts of these minerals have been utilised in the manufacturing industries.

Examination of dust extracted from the lungs of individuals employed in the production of kaolinite have shown that large amounts of this mineral can be retained in the lungs. Kaolinite particles are inhaled as aggregates of kaolinite crystals which remain unaltered once they have been retained in the lungs. They may be associated with small quantities of mica and anatase, with minor amounts of quartz, all of which can also be identified in the dust retained in the lung.

The morphology of the kaolinite particles observed is such that the hexagonal form of the mineral particles can readily be detected, as illustrated by the electron micrograph in Fig. 7.1d. Kaolinite particles observed in the lungs of occupational exposure cases differ from those observed in environmental exposure cases where the kaolinite particles are in the main irregular, smaller in size, and consist of randomly orientated platelet aggregates.

The examination of lungs of individuals exposed mainly to mica have shown that the morphology of particles is related to the geological source and origin of the mineral. Muscovite derived from primary deposits such as granites produce dust particles which consist of large thin platelets of obvious crystalline material. In the literature it is disease cases arising from exposure to muscovite in this form that has mainly been reported (Davies and Cotton 1983). Examination of cases of exposure to mica arising from sedimentary sources has produced dust with a very different morphology. It consists of particles comprised of aggregates made up of irregular platelets which are sub-micron in size. These dust particles can be more accurately described as illites or hydromuscovite particles, and they are illustrated by Fig. 7.1c.

The dust retained in lung tissue after long-term exposure to vermiculite has a morphology similar to that of muscovite.

The morphology of montmorillonite dust particles observed in the lung tissue of occupationally exposed cases are illustrated in Fig. 7.1b. They have been found to be similar in their physical characteristics to

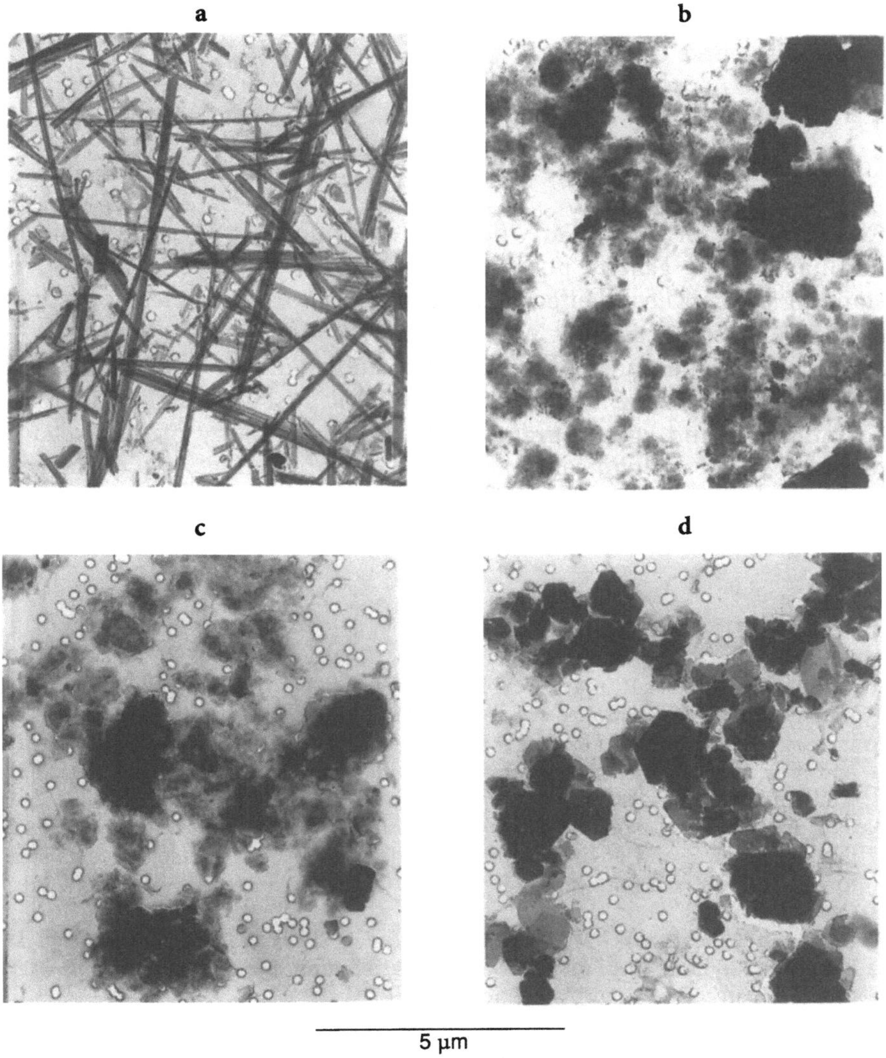

Fig. 7.1. Electron micrographs of respirable clay mineral dusts. **a** Palygorskite; **b** bentonite; **c** muscovite; **d** kaolin, recovered from animal (**a**) and human (**b, c, d**) lung tissue

those of the sedimentary mica minerals consisting of ill-defined aggregates of poorly crystalline particles.

The fibrous clay minerals palygorskite and sepiolite are rarely found other than in the lung tissue of occupationally exposed individuals. The dust particles they form consist of aggregates composed of very fine fibrous particulates. The morphology of these fibrous particles can be similar to those illustrated in Fig. 7.1a, although their dimensions vary with their source of origin.

7.6 Conclusion

Clay minerals, in general, are the most common dust particles found in the lungs of the general population on a worldwide basis. Our understanding of their health effects has been formed from studies of occupationally-exposed cases only. With the growing interest now being expressed in our surroundings, more detailed studies of our environment are being undertaken. The amount of atmospheric dust generated by increased industrial and agricultural activity in the aetiology of dust related diseases will be a significant part of this research effort. The qualitative and quantitative assessment of atmospheric dust and its biological potential will be a very important part of future integrated pollution control.

References

Amandus HE, Wheeler R (1987) Morbidity and mortality of vermiculite miners and millers exposed to tremolite-actinolite. Part II. Mortality. Am J Ind Med 11:15–26
Amandus HE, Althouse R, Morgan WKC, Sargent EN, Jones R (1987) The morbidity and mortality of vermiculite miners and millers exposed to tremolite-actinolite. Am J Ind Med 11:1–37
Baris I, Simonato L, Artvinli M, Pooley FD, Saracci R, Skidmore J, Wagner JC (1987) Epidemiological and environmental evidence of the health effects of exposure to erionite fibres: a four year study in the Cappadocian region of Turkey. Int J Cancer 39:10–17
Begin R, Rola-Pleszcynski M, Masse S, Lemaire I, Sirois P, Bocter M, Nadeau D, Drapeau G, Bureau MA (1983) Asbestos induced lung injury in the sheep model: the initial alveolitis. Environ Res 30:195–210
Davies D, Cotton R (1983) Mic pneumoconiosis. Br J Ind Med 40:22–27
Fennerty A, Hunter AM, Smith AP, Pooley FD (1983) Silicosis in a Pakistani farmer. Br Med J 287:648–649
Gamble J, Sieber WK, Wheeler RW, Reger R, Hall B (1988) A cross-sectional study of US attapulgite workers. Ann Occup Hyg 32:475–481
Gibbs AR (1990) Human pathology of kaolin and mica pneumoconiosis. In: Bignon J (ed) Health related effects of phyllosilicates. NATO ASI Series, G 21. Springer, Berlin Heidelberg New York, pp 217–226
Gibbs AR, Pooley FD (1994) Fuller's earth (montmorillonite) pneumoconiosis. Occup Environ Med 51:644–646
Gibbs AR, Pooley FD, Griffiths DM, Mitha R, Craighead JE, Ruttner JR (1992) Talc Pneumoconiosis. Hum Pathol 23:1344–1353
Harington JS, Miller K, McNab G (1971) Haemolysis by asbestos. Environ Res 4:95–117

Hessel PA, Sluis-Cremer GK (1989) X-ray findings, lung function and respiratory symptoms in black South African vermiculite workers. Am J Ind Med 15:21–29
Hunter D (1994) Hunter's diseases of occupations. In: Raffle PAB, Adams PH, Baxter PJ, Lee WR (eds) The diseases of occupation. Edward Arnold, London
IARC Monographs on the evaluation of carcinogenic risks of chemicals to humans. Silica and some silicates (1987) No 42. International Agency for Research on Cancer, Lyon
International Labour Office (1980) Guidelines for the use of ILO international classification of radiographs of pneumoconiosis. Occupational Safety and Health Series, No 22, ILO, Geneva
Jaurand MC, Pairon JC (1990) Review of short-term tests used for the toxicological evaluation of phyllosilicates. In: Bignon J (ed) Health related effects of phyllosilicates. NATO ASI Series, G 21. Springer, Berlin Heidelberg New York, pp 331–347
Kennedy T, Rawlings W, Baser M, Tockman M (1983) Pneumoconiosis in Georgia kaolin workers. Am Rev Respir Dis 127:215–220
Landas SK, Schwartz DA (1991) Mica-associated pulmonary interstitial fibrosis. Am Rev Respir Dis 144:718–721
Lapenas D, Gale P, Kennedy T (1984) Kaolin pneumoconiosis: radiologic, pathologic and mineralogic findings. Am Rev Respir Dis 130:282–288
McConnochie K, Bevan C, Newcombe RG, Lyons JP, Skidmore JW, Wagner JC (1993) A study of Spanish sepiolite workers. Thorax 48:370–374
McDonald JC, McDonald AD, Armstrong B, Sebastien P (1986) Cohort study of mortality of vermiculite workers exposed to tremolite. Br J Ind Med 43:436–444
McDonald JC, McDonald AD, Sebastien P, Moy K (1988) Health of vermiculite miners exposed to trace amounts of fibrous tremolite. Br J Ind Med 45:630–634
Nagleschmidt G (1956) Inter-laboratory trials on the determination of quartz in dusts of respirable size. Analyst 81:210–215
Norboo T, Angchuk PT, Yahya M, Kamat SR, Pooley FD, Carrin B, Kerr IH, Bruce N, Ball KP (1991) Silicosis in a Himalayan village population: role of environmental dust. Thorax 46:341–343
Parkes WR (1994) Occupational lung disorders, 3rd edn. Butterworths, London
Pooley FD (1977) The use of an analytical electron microscope in the analysis of mineral dusts. Philos Trans R Soc Lond A 286:625–638
Pooley FD (1981) Tissue mineral identification. In: Occupational lung diseases: research approaches and methods, chap 10. Marcel Dekker, New York, pp 189–235
Pott F, Bellman B, Muhle H, Rodelsperger K, Ripe RM (1990) Intraperitoneal injection studies for the evaluation of the carcinogenicity of fibrous phyllosilicates. In: Bignon J (ed) NATO ASI Series, G 21. Health related effects of phyllosilicates. Springer, Berlin Heidelberg New York, pp 319–329
Rundle EM, Sugar ET, Ogle CJ (1993) Analyses of the 1990 chest health survey of china clay workers. Br J Ind Med 50:913–919
Sahu AP (1990) Biological effects of mica dust on experimental animals. In: Bignon J (ed) NATO ASI Series, G 21. Health related effects of phyllosilicates. Springer, Berlin Heidelberg New York, pp 386–393
Sakula A (1961) Pneumoconiosis due to Fuller's earth. Thorax 16:176–179
Sheers G (1964) Prevalence of pneumoconiosis in Cornish kaolin workers. Br J Ind Med 21:218–225
Stanton MF, Layard M, Tegeris A, Miller E, May M, Morgan E, Smith A (1981) Relation of particle dimensions to carcinogenecity in amphibole asbestoses and other fibrous minerals. J Natl Cancer Inst 67:965–975

Vallyathan V, Schwegler D, Reasor M, Stettler I, Clere J, Green FHY (1988) Comparative in vitro cytotoxicity and related pathogenicity of mineral dusts. Ann Occup Hyg (Suppl 1), 32:279–287
Wagner JC (1990) Review on pulmonary effects of phyllosilicates after inhalation. In: Bignon J (ed) NATO ASI Series, G 21. Health related effects of phyllosilicates. Springer, Berlin Heidelberg New York, pp 309–318
Wagner JC, Sleggs CA, Marchand P (1960) Diffuse pleural mesothelioma and asbestos exposure in the North Western Cape Province. Br J Ind Med 17:260–271
Wagner JC, Berry G, Timbrell V (1973) Mesotheliomas in rats after inoculation with asbestos and other materials. Br J Cancer 23:173–185
Wagner JC, Berry G, Skidmore JW, Timbrell V (1974) The effects of the inhalation of asbestos in rats. Br J Cancer 29:252–269
Wagner JC, Skidmore JW, Hill RJ, Griffiths DM (1985) Erionite exposure and mesotheliomas in rats. Br J Cancer 51:727–730
Wagner JC, Pooley FD, Gibbs AR, Lyons J, Sheers G, Moncrieff CB (1986) Introduction of china stone and china clay dusts: relationship between the mineralogy of the dust retained in the lungs and pathological changes. Thorax 41:190–196
Wagner JC, Grifiths DM, Munday DE (1987) Experimental studies with palygorskite dusts. Br J Ind Med 44:749–753
Wastiaux A, Daniel H (1990) Pulmonary toxicity of kaolin in rats exposed by inhalation. In: Bignon J (ed) NATO ASI Series, G 21. Health related effects of phyllosilicates. Springer, Berlin Heidelberg New York, pp 405–414
Zaidi SH (1969) Experimental pneumoconiosis. The John Hopkins Press, Baltimore

Glossary

adenocarcinoma	a tumor which form glands and frequently secretes mucous substances
alveolus	an air sac where gas exchange takes place
bronchioles	an airway internal to the lung which is smaller than a bronchus and does not contain cartilage in its wall
bronchus	an airway which leads to or is within the lung and which contains cartilage in its wall
caesarean	an operation on the uterus which allows the removal of a foetus through the abdomen
carcinogenicity	properties of a substance which can lead to the development of a neoplasm (tumour/cancer)
carcinomas	a malignant tumour generally derived from epithelium
cytotoxic	toxic to cells
epidemiology	the study of the distribution and causes of disease in human populations
fibrosis	scarring
fibrogenic	cause scarring
granulomas	focal collections of macrophages and giant cells

haemolytic	lysis of red blood cells
hyperplasia	increase in number of cells
lymph	fluid contained within lymphatic vessels
lymphoid	pertaining to the lymphatic system or cells such as lymphocytes, plasma cells
lysozomal	a cellular organelle containing enzymes
macrophage	a scavenger cell
malignancy	malignant tumour
malignant	tending to produce death or disease by infiltration of tissues and/or metastasis
mesotheliomas	tumours derived from the mesothelium which lines cavities such as the pleura, peritoneum and pericardium
morbidity	illness but not death
mortality	death
necropsy	post mortem
neoplastic	a new growth of tissue (tumour) which can be benign or malignant
pathogen	an agent that can cause disease
pathology	the study of disease
peritoneum	the tissue lining of the abdomen and abdominal organs
pleural	the tissue lining of the chest wall and lungs
pneumoconiosis	diseases of the lungs caused by the inhalation of mineral particles
pulmonary	pertaining to the lung
radiograph	X-ray
respiratory system	collection of organs pertaining to respiration (breathing) which includes nose, nasal passages, larynx, trachea, bronchus, bronchioles and lungs
reticulin	protein substance similar to collagen, a constituent of connective tissues
sarcoma	malignant tumour derived from soft tissues such as fibrous tissue, muscle, bone etc.
trachea	main tube connecting larynx to large bronchi which conducts air to and from the lungs

Subject Index

MIX
Papier aus verantwortungsvollen Quellen
Paper from responsible sources
FSC® C105338
FSC
www.fsc.org

If you have any concerns about our products,
you can contact us on
ProductSafety@springernature.com

In case Publisher is established outside the EU,
the EU authorized representative is:
Springer Nature Customer Service Center GmbH
Europaplatz 3, 69115 Heidelberg, Germany

Printed by Libri Plureos GmbH
in Hamburg, Germany